MW00451254

Jean Heidmann is an astronomer at the Paris Observatory who specializes in the search for advanced forms of life in space. Most of his research work as a radio astronomer has been on the properties of galaxies and in cosmology. In the last dozen years he has applied this background to the questions of whether intelligence exists elsewhere in the universe, and, if so, how we can search for it. Heidmann is Secretary of the Bioastronomy Commission of the International Astronomical Union, which is the official body charged with responsibility for extraterrestrial intelligence. He is also a member of the International Academy of Astronautics, and in the role he contributes to the Academy's work on the search for extraterrestrial intelligence. An author of more than 200 research papers, Heidmann has also been Editor-in-Chief of the prestigious research journal *Astronomy and Astrophysics*. He has written several books at a general level, and his cosmology book *Cosmic Odyssey* was translated into five languages.

If extraterrestrial intelligence exists, then positive detection of it would be the greatest scientific discovery of all time. By what criteria should we judge whether we are alone in the cosmos, and how should we set about detecting extraterrestrials? Jean Heidmann answers these questions in this engaging discussion of extraterrestrial intelligence. The author shows how planets fit into the hierachy of the universe, and discuss prebiotic stages of life, and the emergence of primitive biological molecules in the solar system. From this base the entire subject of extraterrestrials is explained: techniques and the results of current projects, the expansion of searches for extraterrestrials, the habitable zones in our universe, and what might happen if actual contact takes place. Our generation is capable, in principle, of communication across interstellar space, bounded only by the speed of light, and soon it will be possible to set tight limits on the presence or absence of extraterrestrials in our Galaxy.

Related titles

Remarkable Discoveries
FRANK ASHALL

Hard to Swallow: a brief history of food
RICHARD LACEY

An Inventor in the Garden of Eden
ERIC LAITHWAITE

The Outer Reaches of Life
JOHN POSTGATE

Prometheus Bound: science in a dynamic steady state
JOHN ZIMAN

JEAN HEIDMANN

Translated by Storm Dunlop

Extraterrestrial Intelligence

CAMBRIDGE
UNIVERSITY PRESS

Published by the Press Syndicate of the University of Cambridge
The Pitt Building, Trumpington Street, Cambridge CB2 1RP
40 West 20th Street, New York, NY 10011-4211, USA
10 Stamford Road, Oakleigh, Melbourne 3166, Australia

First published in French in 1992 by Éditions Odile Jacot, Paris,
as *Intelligences Extra-terrestres* and © Éditions Odile Jacot 1992

This edition published in 1995 by Cambridge University Press as
Extraterrestrial Intelligence and © Cambridge University Press 1995

Printed in Great Britain at the University Press, Cambridge

A catalogue record of this book is available from the British Library

Library of Congress cataloguing in publication data

Heidmann, Jean.
[Intelligences extra-terrestres.   English]
Extraterrestrial intelligence / Jean Heidmann : translated by
Storm Dunlop
p.   cm.
ISBN 0 521 45340 2
1. Life on other planets.   2. Exobiology.   I. Title.
QB54.H39613   1995
574.999 – dc20   94-34829 CIP

ISBN 0 521 45340 2 hardback

TAG

Once again, my dear Marie, I dedicate a book to you. After the Scylla and Charybdis of the Cosmic Odyssey we now set sail for distant New Worlds.

Contents

Preface

As we come to the end of the second millennium, our view of the universe has been radically altered, and our perception of the cosmos has expanded. In future, we will see life as a natural phenomenon that has arisen as part of the evolution of the universe as a whole. If this is really true, the wondrous adventure that led to the appearance and subsequent evolution of life could well have occurred elsewhere, not just on Earth. We have thus ceased to see life as an exclusively terrestrial phenomenon, but instead as one that possesses potential on a cosmic scale, and one that we must consider in terms of the universe as a whole. We have moved from the concept of a physical universe to one of a biological universe.

Our study of the origin of life, our exploration of space, and astronomy itself all prompt us to give serious consideration nowadays to the idea of extraterrestrial life, which until now has been the preserve of science fiction. Straightaway, we need to be more precise: the word 'extraterrestrial' naturally evokes images such as those in *The War of the Worlds*, *Alien*, *ET*, or *Close Encounters of the Third Kind*. We tend to think either of gentle beings or of horrible monsters endowed with amazing powers. Above all, we tend to confer on them a degree of intelligence at least equal to our own. In contemporary literature or films, extraterrestrials are generally either an idealization of what humanity would like to be, or else a caricature of what we fear it might become. In particular, when we envisage forms of extraterrestrial life, we tend to ignore the fact that they might not have attained the level of intelligence and civilization that we have reached. Yet, on a cosmic scale, the evolutionary path that life may follow is highly complex. To understand it, we need to forget the little green men that otherwise haunt our imaginations.

The adventure is, moreover, all the more fascinating – because it is true.

Viewed on a cosmic scale, the evolution of life consists of five principal stages:

> A *cosmic stage*, beginning with the Big Bang (when space and matter appeared), followed by the synthesis of the chemical elements such as carbon – which are of fundamental importance for life as we know it – together with the formation of stars and planets.

> An *organic stage* that saw the formation of the first molecules that were to become the basis of our own life forms. These are molecules like those that radio astronomers have found in interstellar space, space probes have detected in comets, and biochemists have discovered in meteorites that have reached the surface of the Earth.

> A *prebiotic stage* that saw the creation of the 'building blocks' of life, which were more complex molecules, but still devoid of life itself. Such molecules include the amino acids, which are essential components of proteins, and the nitrogenous bases that form the 'rungs' of the double helix of DNA. Such a prebiotic chemistry may be taking place today on Titan, the largest of Saturn's satellites.

> A *primitive biological stage*, with life forms like those of bacteria, which ruled our own Earth for the first few billion years, and which astronomers hope to discover, perhaps in a rather different form, in the permafrost layers of Mars.

> And, finally, an *'advanced' stage*, perhaps even more highly evolved than our own. Nothing in our study of the universe suggests that we are the pinnacle of cosmic evolution. Quite the contrary, in fact.

Apart from the exploration of space, sustained by our knowledge of biology, physics and chemistry, the only method at our disposal of detecting advanced life forms that may have developed beyond our own atmosphere is by patiently searching for the radio signals that they may emit. Behind their giant radio telescopes, astronomers have become spies, electronic eavesdroppers ... What is now known as SETI (the Search for Extraterrestrial Intelligence) was born in 1959. More recently, in 1982, the International Astronomical Union set up a Commission devoted to bioastronomy. Above all, however, in 1992, NASA commissioned a wide-ranging program, based on new technology, which was intended to scan tens of millions of channels at radio frequencies between then and the year 2000. Various other countries including France are also involved.

Before we attempt to visualize the day when someone may actually capture an artificial signal coming from space, let us explore both the intellectual implications of this work, and also the very latest developments in bioastronomy, astronomers' search for life in the universe.

FROM THE PHYSICAL WORLD TO THE BIOLOGICAL UNIVERSE

The idea of extraterrestrial intelligence has a long history, because it goes back to the Greek atomists and Aristotle, in the 4th century BC. Revived and given fresh impetus by the work of Copernicus and Galileo, which revealed the similarities between the Earth and the five 'wandering stars' (the planets Mercury, Venus, Mars, Jupiter, and Saturn) that had been known since antiquity, it was well-established among philosophers in the 18th century. In 1859, however, two crucial advances occurred: in that year, Charles Darwin published his *Origin of Species*, and Gustav Kirchoff used spectroscopy to identify chemical elements in the Sun. Thenceforward, everyone knew that life could arise through the processes of physical evolution, and that matter is everywhere the same*.

Jeans' Tidal Theory

Despite these advances, theories developed at the beginning of the 20th century were nearly fatal to the idea of life in the universe. In particular, around 1920, Sir James Jeans developed a theory,

---

* At the Third International Symposium on Bioastronomy, which was held in 1990 at Val Cenis in Haute-Maurienne, France, the astronomer and historian of science Steven J. Dick, of the US Naval Observatory in Washington, gave a fascinating review of this modern concept of the universe.

known as the Tidal (or Catastrophe) Theory, of the origin of the Earth. The latter was thought to have condensed from a stream of material pulled from the Sun by a star that had brushed past it. But such encounters must be extremely rare. As a result, planets and extraterrestrial life would be the exception rather than the rule.

Jeans' eminence was such that his ideas were widely disseminated. In 1943, however, a complete reversal took place: two planetary companions were discovered apparently orbiting the stars 61 Cygni and 70 Ophiuchi. Jeans refined his calculations and had to admit that the 'catastrophe' might have affected one star in every six. Even more significantly, however, Carl von Weizsäcker had proposed a revised form of the theory of a primordial nebula, a concept that may be traced back to the philosopher Immanuel Kant and the astronomer and mathematician Pierre-Simon Laplace. In addition, it was realized that Jeans' theory suffered from a serious fault: an encounter between two stars could not produce planets with nearly circular orbits around the Sun.

The explosion of ideas

Progress by astronomers thus breathed new life into the theory that billions of planets might exist. About the same time (in 1938), Alexander Oparin published his work on the origin of life, Stanley Miller achieved the first synthesis of the building blocks of life, and the first international symposium on the origin of life was held. The stage was set for the arrival of SETI.

In 1959, Giuseppe Cocconi and Philip Morrison published the first scientific study of the possibility of exchanging radio signals over interstellar distances. In the same year, Frank Drake built a special receiver and made the first attempt to monitor two of the closest stars. Since then some 30 years have passed, and SETI, strengthened by the progress made in bioastronomy, is beginning an observational attack on the old question: 'Are we, or are we not, alone in the universe?' In this field perhaps more than any other, science joins forces with some of the most fundamental questions about the What?, the How? and – above all – the Why? of our humble human condition.

FACTS, SPECULATION, FICTION, AND MONEY

Unfortunately, we have only a single example of life, that found on Earth, to guide our search for life in the universe. We can, of course, try to extrapolate from the form we know, which is based on the chemistry of carbon, to other potential pathways. Why not imagine life based on the chemistry of silicon? Similarly, life on Earth depends on chemical reactions that take place in water. Why not consider other solvents?

Such extrapolations may be supported or refuted quite rapidly by undertaking laboratory experiments. In fact, these concepts are not so very far removed from the basic characteristics of our own life forms, in particular from our methods of reproduction or bodily repair, of energy transformation, of sustenance, and of processing information about the external world. They do not suggest any really new or exotic pathways that life might take. Even while considering them, we still retain the model of life as it has developed on Earth.

More daring extrapolations are possible. For example, astronomers have envisaged life based on the neutron-rich material found on neutron stars. It is, in fact, possible to conceive of systems that are, to a greater or lesser extent, capable of sustaining the main functions of life. Such an environment involves energy and particles that are subject to some extraordinary physics, under what quantum physics describes as 'degenerate' conditions. The overall system could give rise to complex structures and to information-processing.

Hoyle's 'Black Cloud'

One of the oldest and, from the point of view of physics, one of the most credible possibilities, was the main theme of *The Black Cloud*, a science-fiction novel written by the famous, and extremely original, British astrophysicist Fred Hoyle. He invented beings that consisted of clouds of magnetized interstellar gas. Flux tubes within the magnetic fields acted as channels for charged particles (electrons and ions), just as blood corpuscles flow through a network of arteries and veins. Information was stored and manipulated as if by a computer, whose electronic components were not solid but consisted of this plasma, frozen in place by magnetic fields. In addition, the clouds could store energy. They led an ideal life, communicating through space by means of radio waves. When their energy reserves began to decline, they propelled themselves into the neighborhood of a star by ejecting a stream of particles. This is how human beings entered the story: one of these clouds had encircled the Sun to capture its energy, spreading panic by obscuring the Sun and causing gravitational perturbations of the Earth's orbit.

Scientists are not prohibited from indulging in outrageous speculations. The latter may, in fact, open up new ideas or points of view. By way of compensation for the aberrations that are only too likely to arise, and which are inimical to knowledge itself, researchers owe it to themselves to suggest, as soon as possible, ways in which their theories may be tested. They need to obtain specific deductions that may be tested observationally.

Dyson spheres

A good example of an attempt at verification has been carried out with respect to the 'Type II' civilizations proposed by Nikolai Kardashev. (These are discussed in detail later, see p. 108.) He described a Type II civilization as one that has been able to master, for its own ends, all the energy produced by its central star. According to the laws of thermodynamics, any civilization of this type would be forced to radiate a substantial fraction of this energy away to space in the form of infrared radiation. According to the cosmologist Freeman Dyson, civilizations of this sort may have been able to use the materials in their asteroids to construct a gigantic shell surrounding their star, enabling them to capture its energy. Such spheres would be infrared sources, and could thus be detected.

Subsequently, IRAS (the Infrared Astronomical Satellite) made a survey of the sky, detecting 130 375 infrared sources that corresponded to stars. From these, Jun Jugaku, Professor of Astronomy at Tokai University in Japan, selected 594 stars that resemble our Sun, and which might therefore shelter some form of life. To detect any additional, i.e., artificial, infrared radiation, he needed to compare the measurements with others of the ordinary radiation from the stars themselves, and these had to be taken from another catalog. Once this had been done, only 54 candidates remained. Eventually, only three of these were shown to have an infrared excess.

After closer examination, however, even these proved to be explained by natural causes. Jugaku concluded that there was no evidence for a Dyson sphere among the 54 candidate stars. One recommendation that arose from this study, however, was that the ordinary radiation from many more stars should be measured. This would enable the rich resources of the IRAS catalog to be utilized to the full. Kardashev and Dyson's speculations may not have been confirmed, but it would be interesting to continue this line of research, which has, so far, been the subject of relatively little effort.

The question of money

As we can see, searching for life elsewhere in the universe requires the development of new observational and experimental instruments and technologies. It therefore naturally requires finance. Science is generally organized in such a way that the finance comes from governmental sources; sources that are responsible to taxpayers, and which only disburse funds to projects that appear really sound. We may conclude that because it is difficult to obtain funds for even the most rigorous scientific projects, suggesting research programs that are based on speculation – far less on fiction – is out of the question.

As a result, in searching for life in the universe, scientists must turn to facts for support. Unfortunately, once again we have only one example to point to: that of life on Earth.

THE EARTH AS AN EXAMPLE
The age of the Earth

The Earth was formed 4 555 000 000 years ago. This is 1000 times as long as the time when the earliest of our own ancestors walked upright. It is a million times as long ago as the first historical civilizations. Its age is known with an accuracy that is astonishing to astrophysicists, who, in general, have to rest content with values accurate to a few tenths, both because it is difficult to wrest data from the cosmos, and also because of problems in devising suitable theoretical frameworks. By way of comparison, the Big Bang cannot, as yet, be dated more accurately than having taken place between 12 billion and 20 billion years ago.

Where measurements concerning our own world are concerned, an accuracy of tenths of 1% is attained, thanks to methods of determining the radioactivity of the oldest atomic nuclei that it contains. Their half-lives are well known, and they are not sensitive to any outside influences. All that we need do is to measure, as accurately as possible – and not without problems – the relative abundances of the nuclei that have been produced by radioactive disintegration and of the original nuclei, and we can establish the age of the Earth.

This corresponds to the period when our planet became an individual condensation, isolated from the rest of the primordial nebula; in other words, when practically all the Earth's mass had gathered together into a relatively dense sphere. It is estimated that between the beginning and the end of the condensation phase (that is, between the collapse of the protosolar nebula and the agglomeration of the Earth), about 100 million years elapsed, which is a fairly short time on a cosmic scale.

The primitive stage

When it was born, the Earth was completely molten because of the heat produced by the impact of planetesimals (primitive, small planetary bodies), cometary nuclei, and clumps of dust and gas, all of which were attracted by the young Earth's force of gravity. This molten state led to the globe becoming differentiated: the heavy elements, mainly iron and nickel, gathered in the center, where they formed the core, around which there was a thick layer of primitive magma. The thick primordial atmosphere that surrounded the globe arose from degassing of the molten material. It was similar to the gases ejected by volcanic eruptions today, and consisted of carbon dioxide, nitrogen, and other, more complex, molecules, such as methane and sulfuric acid.

Most of this primordial atmosphere was blown away into space by the powerful 'wind' arising from the violent activity undergone by the Sun when it passed through the temporary T-Tauri stage of its evolution (see p. 12). It is possible that later, during the next 100 million years, a second primitive atmosphere, consisting mainly of water vapor and carbon dioxide, may have arisen through the impact of comets that originated beyond the orbit of Jupiter.

Subsequently, the globe began to cool; silicates rose to the surface of the magma and began to solidify. The oldest rafts of granite that are known – those forming the base of the Canadian Shield, for example – are 3.8 billion years old. The atmosphere cooled sufficiently for the water vapor to condense into liquid droplets. Torrential rain began to fall, and some geophysicists estimate that it lasted, without stopping, for 10 million years. If there had been any microscopic ancestors of the human race to witness this torrential rain, it could well have been the basis for the traditional myth of the Deluge.

The ocean planet

Nevertheless, this unique rainfall did leave a souvenir, or rather a present, behind it. Thanks to its existence the atmosphere lost nearly all its carbon dioxide. Without it, the gas, which was extremely abundant, would have blanketed the Earth in a dense layer, and the surface pressure would have been around 100 atmospheres (approximately 10^7 Pa), producing a catastrophic greenhouse effect. The Earth would have suffered the same fate as our neighboring sister planet, Venus, where the surface temperature is 450 °C – and we should not be here!

How did it come to bequeath us this precious inheritance? The rain, which was a mixture of water and sulfuric acid, dissolved the calcium from the basalts and granites in the primitive crust. The calcium reacted with the atmospheric carbon dioxide to give calcium carbonate, which was deposited at the bottom of the early oceans, where thicker and thicker carbonate sediments accumulated.

This transformation of atmospheric carbon dioxide into subterranean carbonate rocks changed the Earth's destiny. It became an ocean planet. Its residual atmosphere produced a slight greenhouse effect, far less than that on Venus, but enough to protect the world from the cold of interplanetary space. The Earth was therefore able to press on with the grand adventure of life, because its temperature range remained between 0 and 100 °C, as required for liquid water to exist at normal atmospheric pressure.

Less than 700 million years after its birth, the Earth became more temperate. The early cores of lifeless continents emerged from a global ocean, beneath an atmosphere that consisted largely of nitrogen, with minor contributions from water vapor, carbon dioxide, and methane. Dense plumes of smoke rose into the sky from raging volcanoes, and from giant meteorite impacts.

In fact, the bombardment, which was extremely intense during the Earth's formation, did not cease suddenly: it decayed over several hundreds of millions of years. Even today, we still see shooting stars, and occasional meteorites such as the one that created Meteor Crater in Arizona, and the suspected giant impact at the boundary between the Cretaceous and Tertiary periods, as well as the fragment of a comet or stony asteroid that fell over Tunguska in Siberia in 1908.

The sky above the rocky landscape 3.8 billion years ago was dominated by the Moon that we know today. It would have seemed gigantic, because it was much closer than it is now, and had not been pushed out into space by tidal effects. Despite this, its actual appearance closely resembled the one we know. An observer would have been able to see the occasional impact of one of the bodies during the later stages of the cosmic bombardment: an immense conical halo of molten fragments would suddenly erupt, only to fall back to the surface within a few minutes, leaving a glowing patch on the surface that would slowly die away over the succeeding months.

The first stage of the Earth's history passed in this fashion. This was the cosmic stage, which progressed solely in accordance with the physical laws governing the universe as a whole. It set the stage for the second act: the organic phase. Yet the first act was the most grandiose of all: it involved the universe as a whole.

Acknowledgments

At my invitation the whole manuscript has been read critically by: Nicloe Hallet, assistant engineer at the Paris Observatory; Antoine Heidmann, CNRS research fellow at the École normale supérieure; Marie-Ange Heidmann, lecturer at the Palais de la Découverte, Paris; Monique Ruyssen, my little sister 'Caroline'.

Sections relevant to their particular specialties have been read by: François Biraud, CNRS Research Director at the Meudon Observatory; Lucette Bottinelli, professor at the University of Paris-Sud, Orsay; André Brack, CNRS Research Director at the Center for Molecular Biophysics, Orléans, Alain Cirou, Chief Editor of *Ciel et Espace*; Yves Coppens, professor at the Collège de France; Emmanuel Davoust, astronomer at the Midi-Pyrénées Observatory; Lucienne Gouguenheim, professor at the University of Paris-Sud, Orsay; Anny-Chantal Levasseur-Regourd, professor at the University of Paris VI, Paris; Jean-Pierre Luminet, CNRS research fellow at the Meudon Observatory; Philippe Masson, professor at the University of Paris-Sud, Orsay; Thierry Montmerle, physicist at the Saclay Research Center; François Raulin, professor at the University of Paris-Val de Marne.

It is with great pleasure that I take this opportunity of recording my appreciation of all their comments and suggestions.

I should also like to thank Jean-Luc Fidel, of the publishers Éditions Odile Jacob, Paris, for the great care that he took in reading my original French text and for his helpful suggestions.

Finally I am immensely grateful to Storm Dunlop for his excellent translation and suggestions for updating the text.

PART I

The bioastronomical prospect

1

The cosmic stage

THE BIG BANG, SPACE, AND MATTER

Volcanic eruptions, earthquakes, droughts, floods, cyclones, tornadoes, and tsunamis are all natural catastrophes that terrify the human race. Some people rail against these events and bewail their bad luck. Yet we need to understand a fundamental fact: contrary to the fairly widespread view that has transformed a blind cosmos into Mother Nature, who is all-protecting and benevolent toward humanity, the universe is, to say the least, utterly indifferent to us. In the words of a 16th-century tanka (a Japanese verse-form), we are 'no more than fleeting foam on the surface of a violent sea.' Our story, our adventure, is a sort of cosmic odyssey, frequently titanic in stature, but also occasionally completely trivial. Typhoons and eruptions are merely minor hazards on the terrifying cosmic scene.

In comparison with them, catastrophes of a completely different order could have affected the whole universe. Matter might not have existed. Without matter, how could life have appeared? Even worse, space might not have existed. Without space, how could life be conceived?

The inflationary Big Bang

Such questions arise from the theory of the inflationary Big Bang, which is the result of the alliance between two different scientific disciplines: one is cosmology, which attempts to grasp the history of the universe as a whole over vast scales of space and time; the other is particle physics, which tries to trace to their fundamental roots the properties, nature, and origin of the elementary particles that are the basis of all matter.

This decade-long collaboration between cosmologists and particle physicists has arisen for a very down-to-earth reason: to test their latest developments the physicists need highly energetic particles of the sort that arose only during the first instants of the Big Bang. Rather than constructing accelerators of truly astronomical dimensions (far larger than the Earth), they need to scour the universe for effects that enable them to check their particle physics.

Quantum uncertainty

The basis of elementary-particle physics is quantum theory, founded by Louis de Broglie in 1923 with his concept of wave–particle duality. The behavior of every elementary particle is governed by a wave, which evolves according to certain equations. This wave may be compared with the ripples or waves on the surface of a lake, which propagate in accordance with the laws of the physics of liquids. Hitting the water with a stick at one point creates a change in the level, which propagates outward in circular waves. Such waves may reach an embankment, where they would be reflected, creating interference patterns where they encounter subsequent waves, and so on.

De Broglie's waves are to be understood as a measure of the probability of finding the particle associated with the wave, at a specific point at a specific instant. The stronger the wave, the greater the chance of encountering the particle. At the instant the stick hits the water, the particle would be at that particular point, but subsequently it might be at any point on a circular wave, and later still it is most likely to be where the reflected waves interfere constructively with other ripples.

To common-sense logic, the most surprising fact about this association, or wave–particle duality, is that there is a degree of uncertainty both about a particle's position and about its velocity. In classical mechanics, an electron is at a specific point at a specific instant, and is moving at a specific velocity. According to quantum physics, all that we can state are probabilities: the electron should be within a certain region and will have a range of possible velocities. This is what is known as quantum uncertainty.

For half a century, a debate raged about this strange aspect of nature. Einstein and others thought that our physical theories were concealing certain additional 'hidden' parameters that would eventually be discovered and enable the particles to be tracked down. Others saw this aspect as a fundamental property of nature. It is only in recent years that crucial experiments, such as those carried out by Alain Aspect at the École normale supérieure [the principal French teacher-training academy], have come down decisively in favor of the second view.

Quantum uncertainty is not indeterminate. It is defined by Heisenberg's uncertainty principle: 'the product of the uncertainty in position, times the uncertainty in velocity, is equal to h divided by m' (where m is the mass of the particle, and h is the Planck constant). As a result, the larger the mass, the smaller the uncertainties become, which brings us back to classical physics. As regards h, it is one of our universe's fundamental constants, equivalent in stature to the velocity of light, c, and Newton's gravitational constant, G. In 1899, from these constants, Planck calculated a time (the Planck time), 10^{-43} second, and a length (the Planck length), 10^{-33} cm, which are of fundamental importance in modern cosmology.

According to Einstein's theory of general relativity – the 20th century's other fundamental theory – at a time 10^{-43} second, the observable universe was 10^{-33} cm in diameter. It was therefore small enough to be dominated by quantum physics. It is at this instant and at this size that we lose all trace of the universe's previous history, and the Big Bang's 'time zero' fades into the mists of quantum uncertainty.

Even if we had a coherent theory that unified quantum physics and relativistic physics, it would still be able to provide just a set of probabilities describing these critical stages in the universe's evolution. Here, for example, are two possibilities offered by semiempirical calculations. At some time in the indefinite past, the cosmos may have undergone a contraction phase, ending in a Big Crunch, more or less symmetrical with its current expansion. Moreover, during a period of several Planck times around time zero, it may have undergone several violent oscillations in size. Another possibility is that it did not exist for an indefinite time in the past, but may have appeared several Planck times before time zero, possibly at a very large size, and collapsed in a similar Big Crunch, preceding our current expansion.

The era within a few Planck times of time zero is still very difficult to understand. Some more definite theories have been advanced, such as the Chaotic Big Bang. According to this, the expansion may have taken place at different rates in different directions. A. D. Linde has advanced the concept of self-reproducing universes, or mini-universes, that form a series of 'bubbles' that are interconnected by a tangled mass of tubes, rather like a bowl of soapy foam. Bumps form on these rapidly expanding bubbles, and the bumps themselves swell and produce a second generation of bubbles.

All these theories suggest innumerable universes that are either connected or isolated from one another, each of which has different geometric and physical properties. We are in one of these universes, which has a specific set of properties.

Universes beyond count

The disconcerting result of these considerations is that the creation of our world was little more than chance: a throw of dice. In particular, after 10^{-43} second, the space containing our own cosmos could have been spherical, Euclidean, or hyperbolic; finite or infinite – just to take the principal models of the universe that have preoccupied cosmologists for half a century since the Russian mathematician Alexander A. Friedmann derived them from general relativity.

Among this ill-defined collection of different forms of space, most are inimical to the emergence of life. Some, for example, would quickly collapse back on themselves, ending their career in a Big Crunch well before the few billion years that are required to cover the long road leading to life had elapsed. Only a few, scattered, individual forms of space could possess the properties required for life. If ours is among the select few, we should not be surprised that we are here to see it; there are no witnesses to stillborn and crippled universes.

Grand unified theory

These ideas are very speculative and concern only the immediate period of time after time zero, lasting a few times 10^{-43} second. In contrast, a much more satisfying theory has arisen to describe the history of the cosmos after 10^{-35} second, by which time it was – although it is difficult to realize it – 100 million times as old.

This conception of the universe is based on the application of what is known as 'grand unified theory' (GUT) to the first few instants of the universe's existence. The theory itself is a generalization of the electroweak theory, which unifies two of the fundamental interactions in nature. The first of these is the electromagnetic interaction, which governs the behavior of: charged particles such as electrons and protons; currents, and electric and magnetic fields; and visible, radio, and other electromagnetic waves. The second is the weak nuclear interaction. This applies to certain nuclear decay processes, the simplest of which is that of a free neutron, which decays into a proton and an electron.

This initial effort to unify two of the fundamental interactions of nature has been immensely successful. It predicted the existence of the intermediate Z and W bosons, 100 times more massive than the proton, which were discovered thanks to the giant accelerator at the European Center for Nuclear Research (CERN) in Geneva.

Following on from this success, grand unified theory attempts to incorporate a third interaction, the strong nuclear interaction, which governs the rest of nuclear physics – namely interactions between protons and neutrons in atomic nuclei and during collisions. The theory is based on quantum chromodynamics, which is the physics of quarks, the fundamental sub-atomic particles that make up protons and neutrons.

Grand unified theories predict the existence of X bosons, which have masses that are not merely 100 times that of the proton, but 10^{15} times! Naturally there is not the slightest chance of constructing a super-CERN on a galactic scale to try to discover X bosons. This is why physicists working on grand unified theory have turned to the cosmologists: prior to 10^{-35} second, the cosmos had a temperature higher than 10^{27} kelvins (K), which was high enough to create X bosons. In that case, it might be possible to try to follow their subsequent history, perhaps even down to our own time, and to study them by means of the 'fossils' that they may have left behind them (which may include magnetic monopoles, cosmic strings, domain walls, etc.).

This is the basis of the theoretical model.

Symmetry breaking

As far as the Big Bang is concerned, history began at 10^{-35} second, well after the 10^{-43} second that marked the end of quantum uncertainty. Our current physics does not allow us to tackle that interval, which remains *terra incognita*. During that time and space, the universe was filled with a seething 'magma' of X bosons, quarks and antiquarks, W and Z bosons, and photons, all controlled by the grand unified interaction; there was complete symmetry between the three interactions (electromagnetic, weak nuclear, and strong nuclear), because identical probabilities existed for the formation of the three types of boson (photons, W and Z bosons, and X bosons) as a result of the highly excited state of the 'magma' – which is indicated by its extreme temperature of more than 10^{27} K.

Because of the expansion, however, the temperature dropped below 10^{27} K: X bosons could no longer be created and thus disappeared. The interaction's symmetry was broken, which caused other bosons to appear. These are known as Higgs bosons, and are, it must be admitted, still highly enigmatic. These bosons tend to align themselves 'in parallel' with one another, releasing a colossal amount of energy, rather like the way in which molecules of water align themselves into crystals on freezing, releasing the latent heat of crystallization. A phase transition occurred between grand unified theory's symmetrical state to a state of broken symmetry, similar to the transition between symmetrical liquid water and ice, which has broken symmetry.

But, as with water, the phase transition is not automatic. Supercooling may occur: water may remain liquid below $0\,^\circ$C, eventually turning, all of a sudden, into ice, releasing its latent energy as it does so. For Higgs bosons, it is estimated that 'supercooling' may persist until 10^{-32} second.

The emergence of space

This is where the inflationary period of the expansion intervenes: between 10^{-35} and 10^{-32} second, the latent, unliberated, phase-transition energy causes a catastrophic, exponential expansion of space. At the end of this phase, space will have expanded by an enormous factor – 10^{50}, according to some calculations.

When the 'supercooling' ceases, whole populations of Higgs bosons become aligned 'in parallel', but the directions in which this takes place may vary from one region to another. What occurs is very similar to what happens when supercooled water freezes: the block of ice is a mass of intermingled crystals with every possible orientation. Each crystal is separated from the others by common boundary walls.

The same applies to space: the 'crystals' of space, defined as consisting of a population oriented in a specific manner, are separated from one another by 'domain walls', which are, in fact, defects in the structure of space, where supercooling still persists.

The appearance of matter

According to the theory, the end of supercooling releases a colossal amount of energy, which creates a plethora of all sorts of particles and antiparticles, including quarks and antiquarks. This event could be considered as a second 'cosmic explosion', a form of second Big Bang, marking the appearance of 'true' matter, i.e., matter in its current form.

These astounding developments in physics are still speculative, but, according to Denis Sciama, one of the most prescient of modern cosmologists, 'It is hard to believe that it is *all* wrong and/or misleading ... We witness the birth of new possibilities for our understanding the universe.'

We should not forget that the Big Bang theory is based on the recession of the galaxies, which itself rests on the interpretation of the redshift of lines in their spectra as being caused by the Doppler effect. Observations of external galaxies by Halton Arp, a celebrated observer at Mount Palomar Observatory, now working at the Max

Planck Institute for Physics in Munich, have, however, indicated that certain spectral shifts may not arise from recession.

Although not necessarily calling the Big Bang into question, these observations might indicate that, as far as extragalactic astronomy is concerned, we may still not understand some aspects of the basic physics involved. But caution is essential. If anyone were to ask me my opinion, I would bet 30 to 1 in favor of the Big Bang.

Abnormal spectral shifts are feasible within the confines of general relativity. Jean-Pierre Luminet (who is carrying out research at the Paris Observatory at Meudon), is, for example, investigating the 'ghost images' that connections in the structure of space, such as those known as 'wormholes', might produce.

The Longest Second

As general relativity's 'traditional', much slower, expansion began after the 'second Big Bang', particles and antiparticles combined, and this occurred increasingly as the universe cooled. It is estimated that by the end of 1 second, nearly everything had been annihilated and converted to photons. Here, however, an extraordinary fact played a part: grand unified theories predict a slight asymmetry between quarks and antiquarks in that a slight excess (amounting to about one-billionth) of quarks were created relative to antiquarks. As a result, when the antiquarks had been annihilated by their corresponding quarks, a small population of unpaired quarks remained. We owe the matter that surrounds us today, and thus our very lives, to these miraculously surviving quarks. In this scheme of things, this second of time, between 10^{-32} second and 1 second, played an absolutely fundamental role in our destiny. This is why I frequently call it 'the Longest Second.'

The Grand Design

After the Longest Second, the universe continued on its way, following the hitherto classic scenario determined by general relativity, which has been termed 'the Grand Design.' It contained a dense mixture of photons, protons, neutrons, and electrons, and had a temperature of some 10 000 million K. Ten seconds later it had cooled sufficiently for it to begin a quarter of an hour of intense nuclear reactions, in which protons and neutrons synthesized helium nuclei, amounting to 25 % of the whole.

Then, for 3000 years, nothing interesting occurred, until the temperature fell below 5000 K, when the electrons combined with protons and the helium nuclei to form the first atoms. For tens of millions of years after this, again nothing of note occurred, except that around 10 million years the temperature was a comfortable 20 °C. In contrast, by 100 million years, it is only about −200 °C. Has the universe come to the end of its career? What a sad fate after such a promising start...

PLANETARY SYSTEMS

Imagining that the cosmos might end up as an ever-growing expanse filled with a mixture of hydrogen and helium that becomes more and more rarefied and colder with time, is to forget the role of chance. Thanks to chance, and perhaps to the irregularities present in the 2.7 K cosmic background radiation 300 000 years after the Big Bang that were revealed by the Cosmic Background Explorer satellite (COBE) in 1992, regions with a slightly higher density appeared. Among these, some were sufficiently strong for the fourth fundamental interaction, gravity, to come into play.

How clusters of galaxies, galaxies, and stars came to be formed during the first few hundred million years remains a mystery. The formation much later of stars and possible planets, however, is beginning to be better understood.

Because of gravitation, spheres of gas were formed that collapsed toward their centers and became increasingly massive. They also became hotter and hotter because of the energy released by their collapse, until eventually thermonuclear reactions were triggered. This is how stars, which provide the light and heat on which our life depends, are born. The process of condensation is very complex, and it is only in recent years that, thanks to new forms of instrumentation, it has been revealed, actually taking place before our eyes, in interstellar space.

Giant molecular clouds

Everything begins with giant molecular clouds, which are vast cool clouds, tens of light-years in diameter, with masses equal to a million Suns, and which contain mainly molecular hydrogen and dust. One of the finest examples is W49A. Located in a distant region of the Galaxy, 50 000 light-years away, and completely hidden by dust clouds in the Milky Way, this immense complex has been revealed by radio waves. Massive stars have been formed within it, each some 50 times the mass of the Sun, and surrounded by shells of ionized gas tens of thousands of astronomical units in diameter (1 AU is the mean Earth–Sun distance). These ionized clouds lie in a ring some six light-years in diameter, which is rotating round a denser, central core, which contains some 50 000 solar masses.

Observations in the spectral lines of carbon monoxide, carbon monosulfide, silicon monoxide, ammonia, and formaldehyde have revealed some remarkable motions within the cloud. The central region of the giant cloud is collapsing toward the ring of stars at a velocity of 14 km/second, creating a shock wave that will reach the latter in about a million years.

Within this gigantic stellar factory, magnetic fields must play an important part. Their lines of force compel ionized particles to spiral around them. A bundle of lines of force acts like a pipe, preventing the particles from flowing across them to any significant extent. These magnetic fields must have played a part in the formation of the stars inside the W49A molecular cloud as a group, and in such an organized manner.

Dark clouds and T-Tauri stars

At the opposite extreme to these immense molecular complexes, most ordinary stars appear in Bok globules, which are small, dense clouds of dust and gas, no more than one light-year across. They are totally obscured by dust and only a few stars, hardly equal to the mass of the Sun, form within them. They may be observed, however, thanks to the infrared radiation that penetrates the cocoon of dust during the final stage of their formation: the T-Tauri stage. During this phase, the protostar is subject to violent events that may last 100 000 years. The phenomena involved include a magnetized stellar wind; the presence of an accretion disk created by the collapse of gas and dust around the star toward the equatorial plane; and internal dynamo effects created by the flow of ionized material within the body of the star, provoking activity similar to that of the Sun, but with large spots that may cover as much as a quarter of the surface.

A good example of such a star is HL Tauri, which already has an accretion disk, 1000 AU across, containing about 0.1 solar mass – i.e., about ten times the mass of our planets – in the form of gas and dust particles smaller than one micron (= 1 micrometer or 10^{-6} meters) in size.

It is estimated that about ten million years elapse between the interstellar cloud beginning to contract and the T-Tauri stage: on a cosmic scale, therefore, the formation of these stars is very rapid.

Jets and disks of gas

An intermediate stage is exhibited by some nebulae known as Herbig-Haro objects. The finest example is HH 111, which exhibits two, diametrically opposed jets that have been produced by a star at its center that is in the process of formation. These straight jets of ionized gas are turbulent and moving at supersonic speeds of several hundreds of kilometers per second. When they encounter the interstellar gas they cause it to glow.

But L1551 is even more revealing, because it has proved possible to detect the jets, the accretion disk (which has the shape of a torus), and the circumstellar envelope surrounding the young star. The jets are perpendicular to the plane of the disk, because it is only here, in the direction of the poles, that they can escape, constrained by the overall, dipole magnetic field. The stellar wind has swept the central portions of the disk clear, so that its shape has become a torus.

Thanks to this information, it has been possible to reconstruct events during this crucial stage in the formation of planetary systems. A shock wave struck the L1551 molecular cloud, compressed a portion of it, which then collapsed along the lines of force of its magnetic field, forming a thin disk perpendicular to the field. This disk contracted radially and began to rotate, and the star formed at its center. The young star's violent activity is blowing the material in the cloud away along the field lines, in two opposing jets at right angles to the plane of the disk.

Protoplanetary disks

This has brought us quite naturally to the question of protoplanetary disks. Despite the discovery of some more recent candidates, the best example remains that of Beta Pictoris, discovered accidentally shortly after the infrared satellite IRAS began operation. This flat disk, which we see almost exactly side-on, has a diameter of 1000 AU. It is orbiting a star that is 53 light-years away, which is brighter than the Sun and expending far more energy, and is just a few million years old. The disk contains dust particles a few microns across, which are, therefore, far larger than those found in interstellar space and more closely resemble those that exist in tails of comets. Their total mass is equivalent to that of Jupiter.

The dust consists of particles of clear ice and grains of dark rocky material, either clumped together or fragmented, which means that the particles probably arose from the debris created by collisions between local comets. Gas has also been detected, and its mass amounts to about one-hundredth of that of the Earth. Its composition varies, and this variation may be caused by the sudden evaporation of kilometer-sized blocks of the disk that fall in toward the star.

Close to the center of the disk, a zone some 15 AU in radius is clear of material; it is thought that it has been swept clear by the gravitational effects of a massive body, which might be a planet. The mechanism is probably similar to that operating in the rings of Saturn, whose extent is governed by various 'shepherd' satellites.

Does Beta Pictoris have a planet?

In 1992, Hervé Beust's doctoral thesis strongly reinforced the assumption that planets might exist. Observations carried out over eight years by the group led by Alfred Vidal-Madjar, Research

Director at the Institut d'astrophysique de Paris [Paris Astrophysical Institute], and theoretical calculations developed by another team headed by Thierry Montmerle, a physicist at the Centre d'études de Saclay [Saclay (Nuclear) Research Center], enabled Beust to describe an interesting sequence of events.

Cometary nuclei would be subject to gravitational perturbations by a planet and would be hurled in toward Beta Pictoris, where they would be vaporized by its radiation. The resulting puffs of gas would produce the intriguing spectral variations that are observed, by absorbing light from the star.

When everything is taken into account, the best solution suggests the existence of a planet with a mass closer to that of Jupiter rather than that of the Earth, taking several years to complete an orbit around the star. This orbit is fairly elongated and slightly larger than that of Jupiter.

A special telescope will be used to monitor the sudden appearances of these puffs of gas, and this should allow us to refine our information about the planet responsible, and perhaps discover others.

A similar object, HD 256, has been discovered, but it is four times as far away and will therefore be more difficult to study.

In his thesis, Beust concludes: 'If the similarities between HD 256 and Beta Pictoris are confirmed, we shall be able to state that the latter is not unique, and this will be of fundamental importance for the study of planetary systems in general.'

Everything seems to point to the fact that we are dealing with a true protoplanetary disk where planets may have been formed. There are, however, similar cases of disks around older stars where we know that planets are no longer being formed ...

It should be noted that even if Beta Pictoris has planets, the star will become a red giant and vaporize its planets before any life similar to our own will have had sufficient time to develop. Among the stellar candidates with disks, however, half-a-dozen have ages of between 2000 and 6000 million years. Advanced civilizations could have emerged around them, which is why I have included them on a priority SETI program at Nançay.

The formation of planets

The stage during which planets are actually formed is still poorly understood. Thanks to the exploration by space probes of the planets and their satellites in our Solar System, however, a possible sequence of events has been outlined. The dust clumped together and the less volatile gases either condensed on their surfaces or solidified into individual grains. The exact gases involved would depend on the ambient temperature of the protoplanetary disk, and thus on the distance from the central star. This clumping together would end with the formation of planetesimals, mini-planets about a kilometer across, and of cometary nuclei.

After this very rapid, physical chemistry stage – it lasts about 1000 years – the role of gravitation becomes all important. The planetesimals are deflected by one another, collide, clump together if they encounter one another at low relative velocities, or are fragmented by head-on collisions. The latter are less frequent, however, because the planetesimals tend to orbit in the same direction as the initial rotation of the nebula. Computer simulations show that in about 100 million years – which is also a very short time – the final planets are formed.

Searching for extra-solar planets

Astronomers are determinedly searching for planets orbiting other stars, both to refine their theories of planetary formation and, above all, to know if life similar to that on Earth might have appeared elsewhere in the universe. In 40 years, 100 000 photographs have been taken, not to detect the planets themselves, which would be far too small, but to detect the slight oscillations that planets would impart to their parent stars as they orbited them, because they are actually orbiting their common center of gravity.

Unfortunately, photographic technology is not adequate, and we have had to wait for the introduction, in recent years, of new methods that are 100 times as sensitive. Some Canadian researchers have been able to measure the variations in the velocity of stars along the line of sight, with an unprecedented accuracy for astronomy, amounting to some 10 m/second – which is the speed of a bicycle.

If these variations are really caused by the oscillations that we have mentioned, observations over a period of a few years appear to have produced a few promising candidates, in particular Epsilon Eridani, the first star Frank Drake scanned for SETI in 1960. It may have a planet whose mass is a few times that of Jupiter.

However, a much better candidate was discovered by accident, although the opportunity was seized independently by an American, David Latham of the Harvard-Smithsonian Center for Astrophysics, and a Swiss, Michel Mayor of Geneva Observatory, which gives it a certain value. They discovered that the star HD 116762, lying at a distance of 90 light-years, shows a regular oscillation, which indicates the presence of a planet with 11 times the mass of Jupiter, completing one orbit in 84 days, like Mercury.

An even more astonishing discovery has been made by A. Wolszczan of Pennsylvania State University. After years of ultra-precise timing of the radio pulses received by the giant dish at Arecibo in Puerto Rico from pulsar PSR B1257+12, he was able to follow the wobblings of the star so accurately that they may be explained by the presence of two planets comparable in mass to the Earth (and possibly another with the mass of the Moon), orbiting with periods of one, two, and three months. Although such low-mass planets are extremely interesting from the bioastronomical point of view, it is doubtful how life could emerge in such a strange environment. Nevertheless, this case is very important as far as the broader question of the formation of planetary systems is concerned.

Beta Pictoris, Epsilon Eridani, and HD 116762 are three rising stars that announce a glorious future for SETI!

2

The organic stage

LIFE ON EARTH

Life began through the slow assembly of organic molecules in liquid water. Such molecules are collections of atoms, built mainly around a scaffolding of carbon atoms, each of which has two shells of electrons surrounding a nucleus consisting of six protons and six neutrons. The first two electrons are in the inner shell, and the other four are in the outer. It is worth noting that a carbon nucleus may also contain two additional neutrons, giving rise to the radioactive carbon-14 (^{14}C) isotope, frequently used for dating fossils.

Carbon chemistry

The external electrons enable the carbon atom to form four bonds with other atoms. In the simplest, most typical, and most symmetrical compound, methane, the carbon atom lies at the center of a tetrahedron, the four corners of which are occupied by hydrogen atoms. But bonds may also be formed easily with oxygen and nitrogen. This gives rise to giant organic molecules that may contain 100 000 atoms, spread among chains and rings of carbon atoms. The molecules most representative of terrestrial life are those of DNA, deoxyribonucleic acid.

Out of the range of nearly 100 atoms found in nature, only carbon possesses this extraordinary versatility in forming compounds, and it is this fact that is the basis of the extreme diversity of organic life. Here again, a lucky chance ensured that carbon nuclei were synthesized in sufficient quantity through nucleosynthesis within

19

the interiors of successive generations of stars. Fred Hoyle has shown that if the carbon nucleus were to have a slightly different excitation state, an intermediate step in its synthesis would have failed to occur, and carbon would be an extremely rare element in the universe.

The universal nature of our own atoms

At the Val Cenis symposium, two astronomers from the University of Pennsylvania, R. E. Davies and R. H. Koch, gave a breakdown of the number of atoms to be found in a 70-kg person. Among the 10^{28} atoms, 38 elements have been detected, of which 27 were of primary importance, particularly oxygen, carbon, hydrogen, nitrogen, phosphorus, and potassium. In the universe the most abundant atoms are hydrogen – at 70%, by far the greatest number – followed by oxygen, carbon, nitrogen and iron, which are found at the percentage level. It is very striking how human bodies selectively amass phosphorus and potassium, increasing their concentration by about 1000 times. Elements that are required in just trace quantities, however, such as molybdenum, do not present any particular bottleneck to growth.

The two astronomers estimated that, through supernovae and stellar winds, practically all the stars in our Galaxy that are more than six billion years old have contributed some hydrogen atoms to each one of us. According to them, 10% of our hydrogen atoms come from neighboring galaxies, such as the Andromeda Galaxy. What is more, gamma-ray photons that arrive here from the most distant objects, such as quasars, create protons and antiprotons when they penetrate our atmosphere, and the former are incorporated into our bodies. Because of mixing on a cosmic scale, our organic chemistry does indeed have a universal origin.

There is also a more moderate degree of mixing on the terrestrial scale: because of the way in which the hydrosphere is recycled, our bodies contain an atom of hydrogen from every milligram of living material more than 1000 years old. So, for example, the horse that Vercingetorix rode when the Gauls surrendered Alésia to Caesar in 52 BC, has bequeathed about a billion atoms of hydrogen to each one of us.

The building blocks of life

Having established the ubiquity of the elements required for organic chemistry – without attempting to prejudge the origin of the simplest organic molecules that might have been found on the early Earth – the beginning of the organic stage on Earth is currently seen in terms of Alexander Oparin's concept of a 'primordial soup', reinforced by Stanley Miller's early experiments.

Nearly 40 years ago, the latter simulated the environment of the primitive Earth in the laboratory. It proved to be easy to synthesize organic molecules that might be described as 'the building blocks of life' – such as amino acids (which are essential components of proteins) – in a flask containing water and an atmosphere of methane, ammonia, and hydrogen, that was subjected to electrical discharges. Since that time, dozens of laboratories, using various energy regimes and different mixtures of chemicals, have tried to produce better simulations of the primitive conditions that recent studies have established as being most likely. These have confirmed the experimental results, without, however, being able to establish that they do reflect what actually occurred. As André Brack, Research Director at the CNRS Laboratoire de biophysique moléculaire [Laboratory of Molecular Biophysics] at Orléans has reported, some people have indeed asked 'if the primordial soup was really fit for consumption.'

The oldest fossils

The first fossil traces of living organisms date back to about 3.5 billion years. They are stromatolites found at North Pole, in northwest Australia, in fossil beach lagoons in a volcanic, marine environment rich in sulfates. Stromatolites are cushion-shaped mounds consisting of successive layers of mineral grains that were trapped by encrusting colonies of microorganisms.

Older fossils, from about 3.8 billion years ago, may have been discovered at Isua, in western Greenland, in the oldest known sediments. Their identification, which was based on morphological features, was highly controversial until the recent results obtained by the geologist J. W. Schopf, of the University of California.

Other investigations have been carried out on fossil organic molecules, which are also 3.8 billion years old. In this case, all that remains is a framework of a few carbon rings. The measured isotope ratios, however, appear to arise from the selective effects produced by biological activity. The conclusion to be drawn, according to M. Schidlowski, is that life on Earth appeared very quickly, in less than 700 million years after the planet's formation some 4.5 billion years ago. The ease with which life appeared may well apply elsewhere in the universe, which is encouraging for those thinking of searching for it.

The eukaryotes

Although, one billion years later, the only living organisms were still just bacteria, single-celled life forms, they had already diversified into three major lines. Around 1.4 billion years ago, eukaryotic cells appeared, which were probably symbiotic associations of these individual forms. This major step forward produced cells 1000 times greater in volume, with complex internal machinery: a nucleus containing DNA, mitochondria for respiration, chloroplasts for photosynthesis, Golgi bodies for excretion, ribosomes for protein synthesis, and flagella for mobility.

The Ediacara fauna

With the increase in the oxygen content of the oceans as a result of photosynthesis carried out by primitive biological activity, another important step was taken with the formation of multicellular animals: the Ediacara fauna. Basically, these were soft-bodied marine creatures. Their largely flattened forms enabled them to present a larger surface area, for a given body volume, facilitating exchange of the oxygen dissolved in the water. They dominated the oceans of the world between 670 and 550 million years ago. Four phyla are known, and some species were dome-shaped, like primitive jellyfish, whilst others resembled primitive arthropods or feather-like sea-pens 1 m long.

The Cambrian explosion

Finally, they were succeeded by a veritable explosion of life forms. This was the Cambrian explosion, which coincided with the conquest of the continents. It is thought that the continual rise in the oxygen produced by photosynthesis finally enabled a protective layer of ozone to form. This explosion of life forms is best illustrated by the survival to modern times of some 30 animal phyla, whose basic structures challenge the most imaginative science-fiction. To name but a few of the most familiar, we have sponges, sea-anemones, worms, insects, starfish, octopuses, and the chordates (or vertebrates). We belong to the chordates, along with fish, reptiles, birds, and the mammals, among whom we are, in the final analysis, of no more importance than mice.

Mammals would probably not have reached the top of the evolutionary tree, if – but this is still a theory – an asteroid had not accidentally crashed into the Earth 65 million years ago, leading to the disappearance of half of all marine species, and, more important for us, of the dinosaurs. If not, some of them, such as *Stenonikosaurus inequalis*, might have been able to evolve intelligence in our stead.

Instead it was the mammals, and in particular the large anthropoids who prevailed. But here again evolution was a lottery. The genealogical trees established by DNA studies show that chimpanzees are closer to humans than to orang-utans. It could well have happened that chimpanzees were the ones using radio telescopes, while we were in zoos, howling for recognition of our status as relatively intelligent creatures.

This is, in outline, how life has evolved on Earth, over some four billion years. Its final flowering took place within about one-tenth of the age of the Earth, and reached its peak in the few million years (one-thousandth of the Earth's age) that separate *Australopithecus*, who first walked upright, and *Homo sapiens sapiens*, who has walked on the Moon.

THE INTERSTELLAR TRAVELS
OF A GRAIN OF DUST

The incredible vacuum of space

Although interstellar space contains giant molecular clouds where the densities may reach $10\,000$ atoms/cm^3, most of it resembles a perfect vacuum, because it contains only one atom/cm^3. Physicists attempting to 'evacuate' a chamber in the laboratory can only dream of such a value. At atmospheric pressure, 22.4 litres of hydrogen contain 10^{24} atoms. Even after pumping down to one-billionth of an atmosphere, 100 billion atoms/cm^3 still remain ... In view of the incredible vacuum of interstellar space, one can only marvel at the extraordinary harvest of results that radio astronomers have gleaned over the last 30 years. They have detected 90 different types of molecule and, what is more, most of them are organic. Whatever sort of mechanism could have synthesized such a variety of molecules in such a perfect vacuum?

Interstellar synthesis

In fact, according to a promising line of inquiry initiated by Mayo Greenberg of the University of Leiden, this mechanism might be found in even rarer interstellar objects: tiny grains of silicate material, one-tenth of a micron across. Only one is found within a cube with sides 100 m long! Processes involving such minute particles have been able to produce such results only because they have had the time, because they have had hundreds of millions of years at their disposal. Here again, we find a situation that is typical of the universe: if tiny numbers, close to zero, are multiplied by others that are extremely large, almost infinite, they give a tangible result.

Only 2 % of the matter 'filling' interstellar space comes from supernova explosions; the remaining 98 % is provided by stellar winds, and consists mainly of streams of particles ejected from their atmospheres by the stars' activity, producing circumstellar envelopes that eventually dissipate into space. When stars begin to approach the end of their thermonuclear stage, becoming cool red giants that are distended to the size of planetary orbits, they eject the tiny grains of silicates that I have just mentioned. Before these finally leave the circumstellar envelopes, their surfaces become covered in water, methane, and ammonia ices.

Subsequently, they begin their long interstellar wanderings, and the cold becomes intense. Each grain is, however, bombarded by ultraviolet photons from stars in the Galaxy. These split the molecules of ice into chemical radicals. Water, in particular, forms the OH radical, which is extremely reactive at normal temperatures, but which remains inactive when it is too cold. Nevertheless, over the course of the tens of millions of years that the grain spends drifting in space, the radicals slowly migrate across its surface, with OH being brought close to NH_2, or CH_3. Then the grain drifts toward a star and heats up. Suddenly, the radicals react violently with one another: CH_3 and NH_2 producing methyl amine CH_3NH_2; CH_3 and OH forming methyl alcohol CH_3OH, etc.

Another long period drifting in space occurs, followed by further heating, either near a star, or by simple collision in a molecular cloud. Finally, after hundreds of millions of years, the grain is covered by a coating of organic material, certain molecules of which, such as formaldehyde HCHO or hydrogen cyanide HCN, have interesting properties for prebiotic chemistry and for the synthesis of the building-blocks of life.

It is estimated that, by processes of this type, interstellar space is able to produce a vast quantity of complex organic molecules, and that, in a typical molecular cloud a few light-years in diameter, their mass may amount to as much as the mass of the Sun.

The grain's wanderings come to an end when it is trapped in a protoplanetary cloud in the process of creating a new stellar system. It will soon be incorporated into the nucleus of a comet, and perhaps captured by a primitive planet, where it might help to initiate life. Some workers believe that when our Earth passed through clouds enriched in organic molecules, it could have accumulated as much as a billion tonnes of organic material, which is more than the current biomass.

Polycyclic aromatic hydrocarbons

A few years ago, the largest organic molecules in interstellar space were discovered by Alain Léger of the University of Paris-VII, and Jean-Louis Puget of the Institut de physique spatiale [Institute of Space Physics] at Orsay, from infrared observations made by the IRAS satellite. These were polycyclic aromatic hydrocarbons (PAH) consisting of sheets of interconnected rings of six carbon atoms, surrounded by hydrogen atoms. Two rings form naphthalene, $C_{10}H_8$, and the most complex measured in the infrared in the laboratory is coronene, $C_{24}H_{12}$, with seven rings.

The most astounding fact reported at Val Cenis by Léger is that these graphite-like molecules are extremely abundant, to the extent that 10 % of interstellar carbon is found in these PAH compounds, making the species the most abundant of the free organic molecules in space, 1000 times as abundant as the next species, formaldehyde.

Among the organic molecules discovered recently in interstellar space, carbon chains take pride of place. In addition to hydrocarbons up to C_6H and cyanopolyynes up to $H(C{=}C)_{10}CN$ there are also chains terminated by sulfur (C_3S), silicon (C_4Si), oxygen (C_3O), or hydrogen (H_2C_4), as well as the pure chain C_5.

As for phosphorus, which is so important for DNA, two molecules, PN and CP, have now been discovered. Although most of the phosphorus may be contained within interstellar grains, the major gaseous reservoir is probably in the form of atomic phosphorus.

These continuing discoveries give us a better and better understanding of interstellar cosmic chemistry, which was the very first step in the universe's organic stage.

THE EXPEDITIONS TO COMET HALLEY

The interstellar wanderings of dust grains may come to an end in a cometary nucleus, where the organic compounds are deposited. Comets are therefore of the greatest interest to bioastronomers. For centuries, these bodies were no better than unidentified flying objects. Their sudden apparitions, strange behavior, changing appearance, and capricious brightness, all argued against their being part of the sphere of the stars, which was the abode of fixed, unchanging bodies, symbols of a world whose very essence was superior to that of our imperfect world below. At a pinch they could be assigned to the atmospheric world, like clouds and lightning. Their relative proximity added to their mystery and made their apparitions all the more menacing.

The sphere of the stars lost its inherent superiority when Tycho Brahe observed the first widely recognized new star, the famous supernova of 1572. The sphere of fixed stars was thus shown to be subject to the same types of vicissitude as our world below. Assisted by progress in physics, Newton's theory of universal gravitation arrived to unify the two worlds: the legendary apple and the distant Moon did truly obey the same laws. Thanks to this key, Edmond Halley managed to untangle the details of earlier apparitions, and predicted the return of 'his' comet in 1758. The apparitions of comets no longer presented a dreadful menace, although at its return in 1910, when Comet Halley's tail would sweep over the Earth, thanks to the persistence of folklore, panic did send a shiver down the spine of some of humanity.

The fly-bys

What did humanity require to free it completely from this ancient bugbear? A just reversal of roles: for humanity to make a fly-by of a comet itself! At the following return of Comet Halley, in 1986, six special space probes – a veritable armada for a first attempt – flew past the comet. Two were Soviet, one European, two Japanese and, at a greater distance, an American probe. The European Giotto probe took the honors, passing less than 600 km from the nucleus, revealing the heart of a comet, an exotic body some 10 km across, for the very first time.

The European space probe Giotto

The Giotto probe, a sizeable cylinder about $1\,m^3$ in volume, was named after the Florentine painter, who represented the Star of the Nativity above the manger in the form of a comet. Calculations have enabled this to be identified with Comet Halley at its apparition in 1301.

The cylindrical probe, covered with solar cells and topped by an antenna for communication with the Earth, had as its primary mission the task of seeing what a cometary nucleus might look like.

The camera only just peeped over the edge of the probe, because it was realized that in making such a close approach, the probe would be subject to bombardment by dust particles ejected from the nucleus. So a compromise had to be struck between the engineers and the astronomers: not too close for the former, so that disaster could be avoided; as close as possible for the latter, so that more details could be seen!

The assessment must have been accurate, because it was not until two seconds before the fly-by (at a distance of 600 km) that a clump of dust larger than the others – but still with a mass of only about 1 mg! – destabilized the probe, interrupting the image transmission for 34 minutes. The accuracy of the fly-by also posed extremely difficult astronavigation problems. Coming within 50 km of a body that is between 10 and 20 km across, lying 100 million km from Earth, with a craft that is moving at a speed of tens of kilometers per second – not per hour! – was something that had never been achieved. Indeed, at first we did not even know the position of either Giotto or the comet sufficiently accurately.

As far as the probe's position was concerned, recourse was made to a world-wide network of radio telescopes, specially called into service to feed the signals from the probe into a central computer, which was then able to calculate Giotto's position in space. As regards the comet, the two Soviet Vega probes, which passed 7000 km away a few days in advance, were given the task of reporting where they saw the comet. This happened just in time for orders to be sent instructing Giotto to fire its rocket motors to achieve the required orbital correction. The achievement must be seen as a triumph for space technology and international scientific cooperation.

Giotto's fly-by of the nucleus may be compared with flying over Paris in a probe that crosses the whole of France in 13 seconds, at an altitude of 500 km. The extremely high speed of the fly-by, 70 km/second, arose because of navigational constraints. None of the probes had an enormous velocity relative to the Earth, being only just sufficient to escape from its gravitational field. They were therefore orbiting the Sun at practically the same velocity as the Earth, 30 km/second, and in the same direction. The comet, however, was moving in the opposite direction at about 40 km/second, which accounts for the relative velocity of 70 km/second that applied at encounter.

The American probe

Another of the armada's probes, the American space probe ICE also showed considerable prowess in navigation. In fact, the United States did not schedule a special mission to study the comet, but, doubtless spurred into action by the European, Soviet, and Japanese efforts for this outstanding first, they organized a quite astonishing, last-minute encounter. For three years, one of their scientific satellites, International Sun–Earth Explorer 3 (ISEE-3), subsequently rechristened International Cometary Explorer (ICE), had been orbiting the Earth at a distance of a million kilometers, engaged in studying the Earth's magnetosphere – i.e., the vast, extremely tenuous shell of ionized particles that flow around the network of lines forming the Earth's magnetic field.

Could the probe be sent toward Comet Halley? Naturally satellites have rockets, controlled by radio, that can provide orbital corrections, but their thrust is far too weak to permit such a trip. But the satellite was orbiting in the region occupied by the Moon. In principle, therefore, it would be possible to alter its path sufficiently for it to pass close to the Moon. During this close approach it would be dominated by the Moon's gravitational field, and its path could be drastically altered. This is the principle of gravitational assistance, which has now become common currency in astronavigation. Without it, the Voyager 2 space probe would never have been able to complete the famous Grand Tour, which took it past Jupiter, Saturn, Uranus, and Neptune between 1979 and 1989.

At the right moment, ICE's thrusters gave it a gentle push in the direction of the Moon. Unfortunately, this thrust was still too little for the first gravitational deflection to send it on toward the comet. It was, however, possible to arrange for ICE to return later close to either the Moon or the Earth, under more favorable conditions.

In this way, after seven close encounters with the Moon, one of which practically brushed the surface – the probe's altitude was just 120 km – the gravitational assistance provided enough energy to send the probe out toward the comet ... but not Comet Halley!

Instead it was sent to Comet Giacobini-Zinner, which was easier
to reach. Nevertheless, this meant that the American probe was
the first to pass through the tail of a comet and, later, came close
enough to Comet Halley to penetrate into the zone where the comet
interacts with the solar wind.

Gravitational assistance

To appreciate exactly what effective gravitational assistance
involves, imagine that you are being chased through a plantation
of young trees: to fool your pursuer, when you are within reach of
a tree, you grab hold of the trunk. This produces a violent swing
(a sling-shot effect), which you can halt by letting go of the trunk,
to do the same thing later on with another tree.

The most recent perfect example is given by the Galileo space
probe, which is destined to fly to Jupiter. Because of lack of funds,
there was no launcher powerful enough to send it there directly.
It was therefore sent toward Venus, which is much easier to reach.
Four months later, Venus sent it back toward the Earth, which it
brushed past ten months later, 960 km above the surface, scarcely
twice the altitude of many of our satellites. Even so, the powerful
gravitational thrust that it received was still not enough to carry it
to Jupiter. Two years later, Galileo again passed close to the Earth,
after having obtained the first photograph of an asteroid, Gaspra,
in the meantime. It will finally reach Jupiter, after having spent six
years traveling between the planets!

The probes' results

Between 6 and 14 March 1986, five space probes passed the nucleus
of Comet Halley. Their closest distances were: Giotto, the European probe, 600 km; Vega 1 and 2, the Soviet probes, 7000 km; and
the two Japanese probes, Suisei at 150 000 km, and Sakigake at 7
million km, without forgetting the American probe ICE, which on
28 March, passed by at a distance of 30 million km.

This vast range of distances enabled all the component parts of
the comet to be examined in detail. There is an enormous contrast
in size between the nucleus, which is measured in kilometers, and

the tail, which attains lengths of millions of kilometers. It is only thanks to the empty vacuum of interplanetary space that such a small object can create such a great effect. When a cometary nucleus approaches the Sun, it heats up and the ices and snowflakes of which it consists vaporize and emit jets of water vapor, which escape carrying with them the dust particles that were intermingled with the ice. The gases are pushed away from the Sun by the pressure of the solar wind, whereas the dust grains, which are denser, despite being subject to radiation pressure from sunlight, tend to trail behind along the comet's orbit. These effects give rise to the sometimes highly spectacular appearance offered by a comet with a straight, bluish, plasma tail, and a second, curved, yellowish, dust tail.

The coma

At an intermediate size, the nucleus is surrounded by an atmosphere at a very low temperature, $-200\,°C$, which is known as the coma. This may have a thickness of $100\,000\,km$, and within it the ultraviolet radiation from the Sun causes chemical decomposition to occur. This is how molecules of water are split into the OH radicals that are detected by radio telescopes capable of making measurements at a wavelength of $18\,cm$. The first such detection was made with the radio telescope at Nançay in 1973, in Comet Kohoutek. Tens of tonnes of water, and as much dust, may be ejected from the nucleus every second! Despite this profligate expenditure, the nucleus loses only a few meters of thickness each time it approaches the Sun, and can last for thousands of orbits. Comet Halley will last for another $100\,000$ years ...

Apart from water, the nucleus also emits a small percentage of carbon monoxide. Anny-Chantal Levasseur-Regourd, who is a professor at the University of Paris VI and research director of the Centre National de la Recherche Scientifique (CNRS) [National Center for Scientific Research] aeronomy service at Verrières-le-Buisson, has discovered that carbon monoxide does not originate solely from the nucleus, but also from the coma. The dust particles, whose structure is like that of snowflakes, release complex organic molecules.

Among these, carbon dioxide, hydrogen cyanide, ammonia, methane, and even formaldehyde and methyl cyanide have been detected. Finally, in 1991, the French radio astronomers Jacques Crovisier and Dominique Bockelee Morvan, from the Paris Observatory at Meudon, announced their detection of methanol in Comet Austin. It is precisely these organic molecules that are responsible for the strong interest in comets and the origin of life.

The dust

As far as the dust is concerned, this consists of grains a few microns in size, made of carbon-rich silicates, and which resemble certain meteorites known as carbonaceous chondrites. The presence of even more complex organic molecules is inferred from the existence of 'CHON' (an acronym indicating the detection of individual atoms of hydrogen H, carbon C, nitrogen N, and oxygen O), in proportions corresponding to molecules such as polyoxymethylene ($-CH_2-O-CH_2-O-$), and perhaps adenine and pyrimidines. A line in the infrared indicates the presence of $C-H$ chemical bonds, which some people link with the polycyclic aromatic hydrocarbons that have been discovered in interstellar space.

These identifications of complex molecules are still preliminary, for the simple reason that the probes flew past Comet Halley at the considerable speed of 70 km/second, and any molecule captured by the on-board detectors was irrevocably broken into fragments consisting of individual elements, which are not, themselves, affected by such impacts.

As compensation for this inconvenience (which was inevitable with the first cometary mission), there is the fact that Comet Halley is the best known of all comets, and thus provides the best understanding of their properties.

Comet Halley's nucleus

Although the most spectacular part of a comet is undoubtedly the tail, the most vital portion is the tiny nucleus. Giotto returned our first surprising view of a nucleus. The surface is extremely dark; it reflects only 4% of the incident light from the Sun, no more than coal. It is far darker than the heaps of snow that are sometimes seen along the sides of city streets, contaminated by all sorts of gases and dirt that accumulate at the surface as the heaps melt.

This is an apt comparison, because this is what a cometary nucleus is thought to be like. Under the slow action of sunlight, the dust, which was originally mixed with the mass of fluffy snowflakes, undergoes chemical changes, which cause the particles to darken, and also accumulate at the surface as the snow slowly vaporizes.

There was another surprise: the surface is relatively warm, 50 °C, precisely because of its dark, absorbent coating, which, according to polarization measurements, is porous. This surface acts like a dark blanket that is heated by the Sun. At the same time, it also prevents the nucleus from evaporating too rapidly in the vacuum of space.

Then there is another fact: the nucleus is larger than was previously thought, because its size was estimated from its apparent brightness, assuming that the surface was as white as clean snow. It actually has an elongated, double shape, like a peanut 15 km long and 10 km wide.

A third discovery is that the topography is very varied for such a small body. There are craters 500 m across, hills several hundreds of meters high, valleys, and slopes of as much as 15 %; in short, a very irregular, cratered and rough surface.

Finally, the most striking fact was the revelation of vents through which jets of gas and dust were caught escaping: these are the origin of the coma and the tail.

What can we deduce from these data about the internal structure of the nucleus? Four models are under consideration: the old model of a 'dirty snowball'; a model of an agglomeration of irregular clumps, loosely bound together and with numerous intervening spaces; a fractal model, with a similar structure over a whole range of scales from a kilometer down to a micron; and a model of porous, rocky blocks cemented together with a mixture of snow and dust. For the moment, it is extremely difficult to decide between these proposals.

COMETS AND LIFE

The reason why comets are of such interest to bioastronomers who are studying the origin of life on Earth, and even of life in the universe, is because their origin confers upon them an exceptional role. Comets are the only bodies that, simultaneously: contain primordial elements involved in the formation of the Solar System; have preserved them in a freezer for billions of years; and are almost within scientists' grasp.

The protosolar nebula condensed about 4.5 billion years ago. This interstellar cloud was moving through a galaxy that was already ten billion years old, and which was therefore greatly enriched in the chemical elements that had been synthesized by earlier generations of stars that have since disappeared. Above all, it had been liberally bestrewn with cosmic dust particles that were coated with the organic molecules that serve as a basis for life. As the Sun and planets formed at the center of the nebula, farther out, in the colder regions, icy planetesimals formed from water molecules and the coated interstellar grains. These planetesimals were not able to grow to more than a few kilometers in size, because the density of material in the outer regions of the nebula was too low.

The Oort Cloud

Studies of the orbits of the comets that visit the inner Solar System revealed that a large fraction of them came from enormous distances: halfway to the nearest stars (some 100 000 AU). This prompted an important and fundamental idea. In 1950, Jan Oort, the universally respected Director of the Leiden Observatory in the Netherlands, proposed the existence of a cloud (the Oort Cloud) that acted as an immense 'cold store' where comets have been preserved since their formation. From time to time, the gravitational perturbation by a star that passes fairly close to the Solar System deflects some of these comets, which then plunge into the center of the System and visit us.

By comparing the numbers of cometary apparitions and of probable stellar perturbations, we can estimate the number of comets in the Oort Cloud: we get the fantastic number of one trillion!

The number of comets in the Cloud may seem quite incredible, but nevertheless, the total mass, although subject to dispute, has been estimated to be between 10 and 1000 times the mass of the Earth – not even one-sixth that of Jupiter. This is another striking example of the numbers encountered in astronomy. They may seem absolutely 'astronomical', and yet amount to very little ... In this field of science it is impossible to judge things at first sight; we need to refer to the actual figures, and check carefully the way in which they are being used.

Although certain comets plunge in toward the center of the Solar System, others must also be ejected into interstellar space. By calculating the number of losses, we find that, despite the enormous population of comets, the supply would soon be exhausted. Somehow it must be replenished. This crucial fact has, in the last ten years, led to much more detailed consideration of the Oort Cloud and of the true site of cometary formation.

The place where comets are formed

Comets are unable to condense at such a great distance as that of the Oort Cloud, which has given rise to the idea that there was a basic, formative region around Uranus and Neptune, at about 30 AU, and that gravitational perturbations caused by the planets during the course of their own formation raised the most distant points of the cometary orbits from the Sun (the aphelia) to some 1000 AU. There, they became subject to perturbations by nearby stars (at some 100 000 AU), which increased the nearest points of the orbits to the Sun (the perihelia) until they lay far beyond the orbit of Neptune, thus removing the cometary reservoir from the influence of the planets. However, according to Levasseur-Regourd, comets may have condensed directly outside the region of Uranus and Neptune.

All these details have been worked out by simulations on powerful computers. In addition, other sources of perturbations have been identified apart from stars, such as the giant molecular clouds that are scattered throughout the galactic disk, and this disk itself, by virtue of its overall mass. This appears to be how, over a period of five billion years, the contents of the central reservoir have been literally pumped out to a distance of several tens of thousands of astronomical units, replenishing the Oort Cloud, and from which a few comets dive in toward the Earth, at 1 AU.

To sum up, most comets were formed as planetesimals in the region of Uranus and Neptune, or perhaps even farther out. They then migrated to the giant Oort Cloud, from which a few occasionally plunge in toward the inner Solar System. They are therefore the true repositories of primitive material in our Solar System.

Primordial material

Although, since the formation of the Solar System, these comets have remained almost permanently at the very low temperature prevailing in the Oort Cloud (around 10 K), similar to that in interstellar space, they have still been subject to bombardment by ultraviolet radiation from the Sun and nearby stars, and also by cosmic rays. In addition, they have been irradiated by supernovae that have exploded in the vicinity, and bombarded by interstellar dust. It is thought that these effects have caused 'gardening' of the surface, and have also coated it with a crust of polymer products.

If, one day, someone wants to obtain some truly primitive cometary material, it will be essential to take the samples from beneath the surface of a comet, from the 'sub-soil', which has never been subject to these various and – in the main – poorly penetrating forms of bombardment.

Comets and biological disasters

The amazing dynamic changes to which the cometary population is subject raise a problem that is of the greatest interest to bioastronomers: what effects might the slightest statistical alteration to this delicate sequence of events have upon terrestrial life? If, for example, a star passes by at 3000 AU, what will it do to the Cloud? In fact, it would drill out a tunnel 500 AU in radius through the Cloud, sweeping up the comets, and causing a shower of some billion comets, which would cross the Earth's orbit over a period of a few million years. Naturally, such a close approach by a star is very rare, and happens on average only every few hundred million years.

If it occurs, however, the rate of impacts on our planet becomes 300 times larger ...

Cometary impacts on our planet could have catastrophic climatic consequences for terrestrial life, causing massive extinctions of certain categories of living things. This is perhaps what happened 65 million years ago, when the dinosaurs disappeared.

When it comes to the extinction of life forms, analysis of fossil species appears to indicate periodic extinctions at intervals of 26 million years, which could perhaps be caused by repeated showers of comets. As a possible trigger, it has been proposed that the Sun has a low-luminosity stellar companion (a brown dwarf) – which has been called Nemesis – in orbit around it, far out in the Oort Cloud, with an orbital period of 26 million years. Another suggestion is the existence of an additional planet, Planet X, orbiting at 150 AU. Yet again, we could resort to the oscillation of the Sun from one side to the other of the galactic plane, with a half-period of 32 million years, as it orbits the center of the Galaxy. Every time it crosses the plane, encounters with giant molecular clouds could precipitate an avalanche of comets.

Comets' contribution to life

It is more likely, however, that when the Earth was first formed, direct cometary falls were far more frequent than subsequently. This time there may have been beneficial consequences for life: they may have seeded the early oceans with organic molecules of interest in triggering the prebiotic stage. Some people even go so far as to suggest that the very water in the oceans could have been brought here by comets.

The starting point for this argument is that at the distance from the young Sun – in its T-Tauri stage – where the Earth accreted, the temperature was too high for water, carbon compounds, and even silicates to have condensed: they were too volatile. As the planets formed and their sizes increased, however, their increasing gravitational effects began to attract the small clumps of material and comets that then existed.

Calculations based on the water content of comets, which is a known factor, and the bombardment rate deduced from impacts on the Moon, show that comets from the region of Jupiter and Saturn could have brought to the Earth, once the latter had formed, an amount of water equivalent to several times the volume of the present-day oceans.

In addition, they could have brought enough silicates to form the terrestrial crust and, finally, enough organic material to form a layer more than 1 km thick, not to speak of a thick secondary atmosphere with a surface pressure 300 times our current atmospheric pressure.

These suggestions are quite impressive, and open up a whole new perspective on the prebiotic seeding of a planet with oceans. Naturally, at that period, the Earth was still hot and subject to major tectonic upheavals. There must have been a considerable struggle between the still hostile world and the delicate, volatile material arriving from beyond Jupiter and Saturn.

THE ROSETTA PROJECT

Even before the 1986 space probes had thoroughly investigated Comet Halley from head to tail, a small group of scientists and engineers were considering the future. Their future was very ambitious, and also very distant. Might it be possible to land on the nucleus of a comet, obtain samples of the surface, and bring them back to Earth for detailed analysis?

Navigating a probe to a comet no longer presents insurmountable technical problems. The encounter should take place when the nucleus is quite a long way from the Sun, beyond the orbit of Jupiter, to ensure that its activity is low. Gravitational sling-shot maneuvers, and navigation 'by eye' using on-board cameras and small thruster rockets, would allow us to bring the probe slowly to within a few kilometers.

Choosing a landing site

Because only the nucleus of Comet Halley is known, the comet being visited would be a new world; all we would know in advance would be its size and thus its mass, at the most. It would therefore be necessary to observe all of it to chose a favorable landing site, avoiding craters, crevasses, the gas vents, slopes that are too steep, areas covered with rocks, or even piles of soft snow. The ideal site would be a flat area covered with a crust of compacted dust, cemented together by the action of solar radiation during previous passages of the comet through the inner regions of the Solar System.

The probe would therefore have to be put in orbit around the nucleus. This would be a particularly special orbit, because the gravity created by the fluffy, low-density nucleus of a comet is tiny, one-thousandth of that on Earth. For a probe to orbit a 2-km nucleus at an altitude of 3000 m, the orbital velocity would have to be extremely low, a few meters per second; any faster and the probe would escape into space.

The landing

After the probe had mapped the topography, using its cameras and its ground-sensing and altimetry radars, and sent the results back to Earth by radio, orders would be given for the probe to land at the chosen site. This would have to be done completely automatically, because, with the distances involved, it could not be controlled directly: at least two hours would be required for the signals' round trip. Any form of rebound has to be avoided, because a probe weighing 1 tonne on Earth would be the equivalent of just 1 kg; the slightest difference between the shock-absorbers on the individual landing legs could cause the probe to rock, or even cause it to take off back into space.

The first operation, which would need to be carried out urgently, would be to anchor the probe with a bolt that penetrated 1–2 m into the surface. At the same time, the nature and the thickness of the crust, which would already have been estimated from the radar observations, would be determined. Echo-sounding would detect the presence of possible cavities that would ruin a core, or of solid rocks that could damage the drill.

Taking samples

Finally, the coring tool would begin to bite into the nucleus, obtaining three successive samples, each 1 m long, which would be hastily stored in the return capsule. Here again, the operation would be quite taxing; it would have to be monitored from Earth, with both the transmissions of the measurements of compaction and hardness, and the commands being subject to long round-trip delays. In addition, the motion of cometary nuclei is complex, with both rotation and precession, like the motion of a top that is slowing down. When the probe is on the opposite side to the Earth, communications would be interrupted. All told, sampling would probably take one or two weeks.

Returning the samples

Automatic return of astronomical samples came of age, technologically speaking, when the Soviet Lunakhod probes, which landed on the Moon in the 1970s, returned a few hundred grams of rocks to Earth. With a cometary nucleus, however, an important difficulty arises: the samples are 'dirty snow' at a very low temperature. It is essential that during the time it takes for them to arrive at our laboratories they are maintained at their original temperature – at least to within a few degrees. This will require efficient cryogenic systems that will be able not only to mitigate the effects of re-entry into the Earth's atmosphere, but also to keep the samples at an appropriate temperature during the long return journey, which may last half-a-dozen years.

Imagine the disappointment and the cries of rage if, when the precious containers are opened, all that is heard is the hiss of escaping vapor from what turn out to be empty tubes – after having spent a billion dollars, ten years on research, another ten years on construction, and a further ten years on the mission, all to wrest the secret of the origin of the Earth and of life from one of the most amazing objects to be found in space!

An engineer's nightmare and a scientist's dream

The dread of such an error is enough to give nightmares to the engineers involved, because politicians and financiers have no desire to run any major risks. In addition, the uncertainties within which the engineers need to work must be kept within bounds. To do so, given that the field is so new and poorly understood, they insist that the scientists provide specific environmental and operational constraints that can guide them in developing the equipment. The scientists, for their part, know just one cometary nucleus (and even that only from a distance), and therefore have great difficulty in satisfying the engineers' demands. This is the atmosphere (albeit a very cordial and friendly one) that prevails during the yearly meetings – generally held in surroundings conducive to peaceful contemplation – where about 100 people discuss these futuristic projects that may one day become reality.

To help the engineers, scientists conceive new methods of observation and for carrying out simulations within the laboratory. For example, the Giotto probe, which survived the Comet Halley encounter with about 60% of its instrumentation in working order, was maneuvered to visit another comet, Comet Grigg-Skjellerup.

A remotely controlled firing of its rockets brought it back, after four years' hibernation, to a gravity assist encounter with the Earth in July 1990, and it finally passed 1000 km from the second comet on 10 July 1992. This is an astronautical record: Giotto is the first probe, and, for the foreseeable future, the only one to encounter two cometary nuclei.

In another approach, several laboratories have tried to simulate cometary surfaces. The largest of these are able to prepare samples 1 m thick. These are created under a vacuum from mixtures of pulverized ice, mineral dust, and organic substances, in various proportions. For several hours, these are held in vacuum chambers and subjected to powerful ultraviolet radiation simulating the radiation from the Sun. Crusts form on the surface, which are measured, analyzed, probed, and punctured, to provide guidelines that can be passed on to the engineers.

In addition, extrapolations are made from the results of these experiments, using computers to calculate as fully as possible the real conditions prevailing in interplanetary space, where long periods of time and enormous dimensions are the rule.

We wish the Rosetta probe luck. Its name, of course, has been chosen because it should allow us to transcribe the information collected from the depths of interstellar space into the same language as the details determined from meteorites. Cometary nuclei are the intermediary slabs of stone that will enable us to read the secrets of the cosmos.

Unfortunately, the unfavorable world-wide economic conditions will probably impose limitations on the project. NASA has withdrawn from the scheme, so the European Space Agency is studying a simplified version, in which just a landing would be made. The investigations would be carried out locally, without necessarily returning any samples. Increased sophistication of robotic and tele-operation techniques would ensure a fruitful mission. The most favored target is Comet Wirtanen, which could be reached in 2011 from a launch in 2003.

According to Levasseur-Regourd:

> The notion of asteroids and comets is outmoded. Certain asteroids are 'defunct' cometary nuclei, and others are still dormant comets. Would it not be better to describe these as comets, reserving the term 'minor planet' for the large asteroids? The classical distinction between asteroid and comet seems to me, from my work on polarization, to be extremely subtle: we are dealing with a continuous transition, rather than two classes of object.

3

The prebiotic stage

Let us now tackle bioastronomy's third stage, the one in which the organic molecules found in the cosmos are converted into the building blocks of life by means of prebiotic chemistry.

TITAN, SATURN'S LARGE SATELLITE

Titan, the largest satellite of Saturn, merits the name. With a diameter of 5140 km, it is second only to Ganymede, 5280 km in diameter, among the planetary satellites, and is slightly less than the 6800 km of the planet Mars and about half the size of our own world. In addition, it is the only satellite to have an atmosphere with a respectable pressure, which at the surface is about one-and-a-half times the atmospheric pressure on Earth.

Finally, and quite exceptionally, this atmosphere consists almost exclusively of nitrogen, as it would be on Earth if biological activity in the form of billions of years of photosynthesis had not accumulated the one part in four that is oxygen and allows us to breathe. Only Triton, the largest satellite of Neptune (which was the last body to be examined by Voyager 2 in 1989), also has an atmosphere of nitrogen, but this is so tenuous that the pressure is measured in microbars (10^{-1} Pa).

The signs of Titan's 'terrestrial' nature do not stop there. In 1980, the American Voyager 1 probe detected hydrogen cyanide. We know from laboratory experiments that it is possible to synthesize guanine and adenine, two of the nitrogenous bases that form the rungs of the double helix of DNA, from five molecules of hydrogen cyanide.

A prebiotic body

This distant body is important for bioastronomy. It is the subject of considerable attention from bioastronomers, to the extent that, in 2004, one of the next great astronomical space probe missions will parachute a probe into its atmosphere, to reveal details of the prebiotic chemistry that may be constructing the building blocks of life in an extraterrestrial environment.

There is, however, one extremely significant difference when compared with the Earth: the temperature at the surface of Titan is −180 °C, which is such a low temperature that any significant chemical reactions must take place extremely slowly. Despite the 4.5 billion years of its existence, Titan has not been able to proceed beyond the prebiotic stage.

But never mind; Titan may be viewed as a model of the primitive Earth held in deep-freeze. This is of considerable interest, because the whole of the prebiotic stage, which is of such fundamental importance in the appearance of life on our planet during the first few hundred million years, has been erased for ever by the intense geological evolution that has completely transformed our planet over the course of time.

Titan therefore offers us a unique opportunity, almost on our doorstep – even though it is some 1500 million km away – to observe a fundamental stage in our own long history, in action, on a planetary scale, and acting over billions of years, rather than over a few weeks in laboratory test-tubes.

The Huygens probe

The project to send a probe into Titan's atmosphere arose at the instigation of Daniel Gautier, Director of the infrared astronomy laboratory at the Paris Observatory at Meudon. This probe has been named Huygens in honour of the Dutch astronomer, who moved to the Paris Observatory at Colbert's request, and who subsequently discovered the true nature of Saturn's rings in 1655.

The probe, which is being constructed by the European Space Agency, will be carried to Saturn by the Cassini spacecraft – named after the first director of the Paris Observatory – being developed by NASA. The composite Cassini–Huygens spacecraft will be launched in 1997, and, after having brushed past Venus, the Earth, and Jupiter to gain gravitational assistance, will arrive at Saturn in 2004, where it will release the probe.

The latter will approach Titan at 6 km/second, and then after initial braking in the upper atmosphere, using a heat-shield, will deploy a small, drogue parachute, and then a larger one. The heat-shield and parachutes will be released at an altitude of 40 km, and the probe will float beneath a third, smaller parachute. For two and a half hours until it reaches the surface, its 40 kg of instruments will analyze the atmospheric chemistry using gas chromatography, mass spectrometry, and pyrolysis, and will also obtain images of the clouds by cameras. For four years, Cassini itself will carry out a series of encounters with Saturn, the rings, and the satellites, including several dozen passes close to Titan.

An atmosphere of organic compounds

The Voyager space probes just flew past the satellite, whose dense atmosphere prevented any observation of the surface. However, spectral analysis revealed numerous organic molecules: methane (a few per cent), hydrocarbons such as ethane, propane, acetylene, methyl acetylene, nitrogenous compounds such as hydrogen cyanide, cyanoacetylene, and dicyanoacetylene, and finally, carbon monoxide and carbon dioxide, all at concentrations below one part in 100 000.

Probing the atmosphere by radio waves revealed the existence of a troposphere extending to an altitude of 40 km, with clouds of methane and ethane, overlain by a stratosphere, with a pressure of 1 millibar at an altitude of 200 km, in which is suspended a layer of organic compounds with, still higher, yet another layer of aerosol haze, again consisting of organic compounds.

What processes could have given rise to such an interesting structure and composition? In the upper atmosphere, ultraviolet photons from the Sun and electrons trapped in Saturn's magnetosphere split molecules of nitrogen and methane into N, CH and CH_2 radicals, which by recombining and further splitting and combining, produce hydrocarbons and nitrogenous compounds.

Laboratory simulations explain the observed concentrations, and predict the presence of even more complex molecules, such as acetonitrile or acrylonitrile, but in quantities too low to have been detected by the Voyager space probes. As for carbon dioxide, whose abundance is around one part in 100 000, it could be produced from methane and water introduced by meteorites.

Various aerosols are produced as a result of this atmospheric chemistry. In the upper stratosphere, acetylene and hydrogen cyanide produce polymers about a third of a micron long, which slowly float downward, and through collisions with one another, aggregate into nuclei about a micron across. Lower down, condensation covers these nuclei with a crust of isobutane. As they drift down toward the tropopause, they are covered with coatings of lighter and lighter compounds (propane, acetylene, and ethane) like the layers of an onion. It is these aerosols that are responsible for the two layers observed in Titan's atmosphere.

When they arrive at the troposphere, the aerosols serve as seeds for the formation of methane, ethane, and nitrogen clouds. Between 40 and 20 km, these clouds consist of crystals about 1 mm across, like our water-ice cirrus clouds. Between 20 and 3 km they are liquid droplets, like our rain clouds. As they fall lower and lower they evaporate and become smaller and smaller, and by the time they reach the surface are reduced to a fine drizzle of almost pure ethane, which is the least volatile compound. This exotic meteorology would produce 300 km^3 of precipitation annually on the surface of Titan, or, to be precise, about one-thousandth of terrestrial rainfall.

The surface

On what sort of soil does this precipitation fall? Although the Voyagers were not able to see anything, Titan is believed to be covered by a vast ocean. The destructive action of ultraviolet radiation from the Sun on the methane in the upper atmosphere would, in fact, lead to its total disappearance in a few tens of millions of years, i.e., in only one-hundredth of Titan's actual age. There must, therefore, be a significant reservoir of methane somewhere. Taken together with the temperature and the surface pressure, this suggests that there is an ocean of liquid methane.

In addition, the chemical processes taking place in the atmosphere produce ethane to the extent that, since Titan's formation, a layer of liquid ethane, 600 m thick, must have been deposited on the surface. What we may then have is an ocean consisting of a mixture of methane and ethane, which has collected the various organic compounds brought down by the rain. The hydrocarbons will have dissolved in this ocean, whereas acetylene and the nitriles, which are poorly soluble, will have been deposited at the bottom in a sedimentary layer, hundreds of meters thick.

We know nothing about the surface relief of Titan, however, so it is quite possible that islands and continents may rise above this ocean. These will probably consist of water-ice. On them, the organic compounds deposited over the course of billions of years may have formed a crust of porous debris, whose cavities may also be filled with liquid methane.

In the face of this mystery, attempts have been made, since the Voyager probes flew by, to bounce radar signals off Titan. (It was just such a method, using the Arecibo radio telescope, that first enabled us to determine the topography and reflectivity of Venus.) In the case of Titan, however, which is much farther away, the task is far more difficult. It proved to be necessary to send the pulses from the Arecibo radar, and use the Very Large Array (VLA) in New Mexico to detect the echoes. Duane O. Muhleman of the Jet Propulsion Laboratory, Pasadena, California, has observed different reflectivities at different points on the surface, which may correspond to oceanic or continental regions.

The fourth planetary robot lander

Uncertainties about the surface will certainly be reduced by the Cassini–Huygens mission, but, in view of the opportunity presented, it has been decided to suspend a small lander beneath the parachute probe. This lander is charged with sending back information directly from the surface. This 'surface scientific package', which weighs 3 kg and has 10 watts of power, is expected to function for a minimum of three minutes.

Developed by the team led by J. C. Zarnecki at the University of Kent, it contains seven, very simple experiments, which might provide unique results. If it falls into an ocean, it will float. Apart from elementary physical and chemical data such as temperature, density, conductivity, and refractive index, it will also measure the speed of sound, and will be able to determine the depth of liquid by sonar. It will also carry an accelerometer and an inclinometer, which, by measuring the waves, will provide new information about the dynamics of the ocean and atmospheric system.

This 'last-minute' package, reluctantly accepted by the engineers, who are always cautious, will be the first robot to land on a fourth celestial body, following the Moon, Venus, and Mars. Will it open the doors to methane tankers sent by the oil companies of the 22nd century?

Prebiotic chemistry

In any case, at present, scientists' interest in Titan lies in the prebiotic chemistry, the third fundamental stage on the path to life in the universe, which may be taking place there. The study of prebiotic chemistry began with the famous experiment carried out by Harold Urey and Stanley Miller in 1953. Miller produced electrical sparks inside a glass flask filled with a mixture of methane, ammonia and hydrogen, whilst simultaneously passing a stream of water vapor through the vessel. He passed the stream of gas through a cold trap, and from the condensate, obtained complex organic molecules, in particular amino acids.

In recent, highly elaborate experiments of the same type, intended to simulate the atmosphere of Titan, Carl Sagan, Director of the Laboratory for Planetary Studies at Cornell University, has identified 59 different species among the gaseous products that are produced, including 27 nitriles. In addition, he has obtained a thick, tarry deposit, brown in color, that he has called a 'tholin', from the Greek word for 'muddy'; its analysis poses the same sort of challenge as that of the organic compounds found in carbonaceous meteorites. Despite this, Sagan has identified polyenes, polycyclic aromatic hydrocarbons, and both biological and nonbiological amino acids.

Prebiotic chemical pathways

According to numerous laboratory exercises carried out since then, in particular those by François Raulin, a professor at the University of Paris-12, Val de Marne, prebiotic chemistry on a planetary scale may be divided into two stages. In the first, reactive organic molecules form in an atmosphere: these are nitriles RCN, and aldehydes RCHO, where R is a radical. They are best produced in an atmosphere containing methane, nitrogen, and water vapor, using energy provided either by ultraviolet radiation or electrical discharges.

In the second stage, these atmospheric precursors further evolve in water – the famous primordial soup – producing the building blocks of life: amino acids, nitrogenous bases, and sugars. The amino acids, in the form $H_2N-RCH-COOH$, take part in the formation of proteins by linking together and eliminating water:

$$\ldots\ HN-RCH-CO-HN-RCH-CO-\ \ldots$$

In life on Earth, these long chains use just 20 different amino acids, corresponding to 20 different radicals R. In meteorites, by contrast, 90 different amino acids have been discovered, of which 8 are identical to ours. This emphasizes both the richness of extraterrestrial organic resources – of natural, not biological origin – and the selectivity of terrestrial life.

The nitrogenous bases, based on a hexagonal ring consisting of four carbon and two nitrogen atoms, arise through polymerization of nitriles in aqueous solution. Hydrogen cyanide, HCN, gives rise to the purines (adenine and guanine) and to the pyrimidines (cytosine, uracil, and thymine), while cyanoacetylene, HC_2CN, leads to cytosine and uracil.

As regards the sugars that are of biological interest, the pentoses, which are built around a pentagon consisting of four carbon atoms and one oxygen atom, can be produced from the aldehydes. In particular, formaldehyde, HCHO gives rise to ribose and deoxyribose.

The sugars, S, and the nitrogenous bases, B, form building blocks, which, together with a phosphoric acid, p (PO_4H_2), make up nucleotides of the form pSB. The latter, by joining together in chains:

can produce the double helix of DNA:

where the two chains are linked by pairs of bases BB, acting as rungs in the double helix. The different types of base B act as the genetic code for all forms of terrestrial life.

Are these prebiotic chemical pathways the ones that prevailed on the primitive Earth? It should be noted that there are considerable difficulties to be surmounted. Taking the primordial soup, for example, if the oceans formed rapidly, the prebiotic compounds would have been diluted very quickly, which would have prevented them from taking part in later reactions. It is possible that lakes or pools might have offered a more favorable environment. The evolution of life may also have arisen through the catalytic properties of certain clays, as suggested by A. G. Cairns-Smith. Such a substrate would mean that molecules formed in an aqueous medium would not be altered by hydrolysis. Another particular difficulty arises from the fact that, when sugars are synthesized under prebiotic conditions, a highly complex mixture is formed of which ribose is only a very small component.

Is such a series of chemical reactions taking place on Titan? Its dense atmosphere contains the necessary ingredients and sources of energy. The first stage involving atmospheric precursors is probably taking place, as indicated by the six hydrocarbons and four nitriles already discovered. As for the evolution of the building-blocks of life, a negative answer is suggested by the absence of liquid water. Any water is frozen and may form continents in an ocean of methane and ethane. Could ammonia dissolved in the ocean serve as a substitute for an aqueous medium? Might the energy injected by cosmic rays into such an ammonia-based system create a pseudo-biochemistry, in which purines, pyrimidines and perhaps even pseudo-polypeptides might be formed? The last word will probably come from the Huygens probe, which, if all goes well, should begin sending back its message at the beginning of the third millennium ...

METEORITES

Radio astronomy, the conquest of space, and macromolecular biochemistry have served as introductions to the broader question of life in the universe. Interstellar molecules, molecular clouds, planetary atmospheres, comets (in general), Comet Halley (in particular) and Titan have all become priority targets in our search for extraterrestrial life. Yet, even on Earth, we are able to lay hands on pieces of the cosmos that have, quite literally, fallen from the sky, and which can be subjected to intense, meticulous scrutiny in our laboratories. Ever since we have recognized their true origin, stony and stony-iron meteorites, bolides and meteors, giant bodies and microscopic stratospheric dust have all brought us into direct contact with extraterrestrial materials.

Even though the intense bombardment that initially shaped our world virtually ceased some four billion years ago, every one of us can bear witness to the fact that some remnants of this activity persist today. Every shooting star is a tiny, belated reminder of our eventful past.

In fact, shooting stars form just part of a much broader class of objects: the meteoroids. These are bodies that orbit in interplanetary space and may encounter the Earth. They give rise to meteors when they enter the upper atmosphere at speeds of 11–70 km/s, leaving a luminous trail behind them at altitudes of between 110 and 70 km.

Sizes

Meteoroids are generally much less than a micron in size, but they may be as large as some 10 km across. Those with a mass less than about one-tenth of a microgram are gently braked above 120 km, without leaving any visible evidence, and they then gently rain down onto the surface. More massive micrometeorites, up to one-hundredth of a gram, cause ionization and leave a trail visible to radar. Between 0.01 g and 1 kg, the trail is visible as a meteor. The lightest may be completely vaporized at altitudes of about 50 km. The others fragment at altitudes between 10 and 30 km, and the pieces then enter free fall: such objects are known as bolides. Finally, those with masses of more than a kilogram – up to a limit of around some trillion tonnes – reach the ground, where they explode, creating impact craters. The debris from bolides and impacts are known as meteorites.

Until about a decade ago, world-wide, we possessed only about 3000 meteorites, which are known by the names of the areas where they fell. The most widely known is the body (with a mass of around 65 000 tonnes), that fell some 40 000 years ago and created Meteor Crater in Arizona, which is 1200 m in diameter. In Australia there is a crater, still clearly visible as a circular rampart of mountains 22 km across, that was formed 130 million years ago. A meteorite weighing several kilograms once fell in a peaceful American family's living room, and as recently as 31 August 1991, an 11-cm meteorite landed with a whistling sound, just 3.5 m away from two boys in Indiana.

A year later, a 12-kg meteorite fell through the trunk of a parked car in Peekskill, NY. What is even more wonderful is that the piece came from a fireball, brighter than the full moon, which traveled some 700 km over West Virginia, breaking into 40 fragments, and that it was videotaped by a dozen people! From these records it was possible to deduce where the object originated. Its orbit stretched over 3 AU, with perihelion close to the orbit of Venus. Before this event, only three other orbits of parent bodies of meteorites had ever been determined.

The Vaca Muerta meteorite

New meteorites are discovered every day. In 1985, for example, astronomers from the European Southern Observatory in Chile found 77 pieces of the Vaca Muerta meteorite on the desert plateau, close to the telescopes. When it fell, some 3500 years ago, the meteoroid was more than a meter across, and had a mass of several tonnes. A metallic fragment weighing about a tonne had been used around 1860 by local South American Indians to make tools, some of which were found in museums.

Naturally, most of this information has been obtained by laboratory studies. This is how we have learnt that the Vaca Muerta meteorite originated in a collision between a minor planet that was partially molten and exhibited volcanic activity, and another minor planet that had a metallic core. Subsequently, the resulting debris from both cooled as a mixture of minerals, half stony and half metallic, which later broke into a stream of fragments, some of which occasionally reach the Earth.

This rare type of 'mesosiderite' (as it is known) has been discovered at only some 30 different sites. In passing, let us admire the painstaking efforts, understanding, and perseverance of the geologists who have succeeded in collecting a few stones with such remarkable histories from various parts of the world.

Types of meteorite

The main types of meteorite are: stones, which consist of silicates; irons, with iron and nickel; and stony-irons, of intermediate composition. Most of the stones are chondrites, containing chondrules, small spherules a few millimeters across, whose mineralogical origin is not at all clear, and which enclose grains of olivine and pyroxene. The chondrites are classified according to the degree of aqueous alteration and thermal metamorphism that they acquired before arriving on the Earth. Finally, there are the carbonaceous chondrites, which are of the utmost importance for bioastronomy. They have a matrix containing carbon, the composition of which is a third characteristic feature. About 5% of meteorites are carbonaceous chondrites.

Most meteorites originate from asteroids (some of which may have reached diameters of 100–400 km), that have been fragmented or partially eroded by mutual collisions. Not only do such meteorites enable us to study fragments of celestial bodies in our laboratories, but some date from the origin of the Solar System. They are extremely precious messengers from the past!

As long ago as the 19th century, the first investigations of the matrix found in carbonaceous chondrites showed that it contained hydrocarbons more or less resembling those found in kerogen, a solid material found in the heaviest oil deposits, such as oil shale. During the years 1950–70, Harold Urey, winner of the Nobel prize for chemistry, working at the University of Chicago, undertook chemical and isotopic analyses that confirmed the existence of aromatic compounds, which were definitely of extraterrestrial origin.

John Cronin, at the University of Arizona, has been responsible for the major developments concerning the presence of amino acids in his favorite meteorite, the Murchison meteorite, which fell in Australia on 28 September 1969. I was lucky enough to be invited, in October 1991, to visit the International School of Space Chemistry, at the Scientific Culture Center in Erice, a wonderful, small, fortified town overlooking the coast of Sicily and the Egadi Islands. It was in this fairy-tale setting that Cronin described his sophisticated analysis of the Murchison meteorite.

Extraterrestrial amino acids

The insoluble part of the matrix contains organic macromolecules, typical of the polymers in kerogen, with structures consisting of a network of aromatic hexagons, interspersed with nitrosyl pentagons, to which are linked COOH and OH functions, and various radicals. In addition, various exotic forms of carbon are present: graphite, silicon carbide SiC, and diamond. The soluble fraction contains silicates and a high proportion of organic compounds. Cronin has identified 74 different amino acids, 87 aromatic hydrocarbons, 140 aliphatic compounds, 10 polar molecules and, above all, the 5 nitrogenous bases found in DNA and RNA! Among the amino acids, 8 reappear in the 20 that are used by terrestrial life in constructing proteins, such as glycine, alanine, valine, and leucine. Some extraterrestrial amino acids, unlike our own, contain as many as eight carbon atoms, have two or three radicals instead of one, a cyclic radical instead of a linear one, or two COOH acid functions instead of one.

According to Cronin, the most striking points are the structural diversity and the fact that every possible compound with between one and five carbon atoms is actually encountered. The overall composition is racemic, that is to say, left- and right-handed stereoisomers are present in comparable amounts, unlike the situation with terrestrial life, which uses mainly a single form. Cronin has also found precursors of these amino acids – i.e., molecules from which they may be derived by chemical reactions – such as the carboxamides:

$$R - C - NH_2$$
$$\|$$
$$O$$

In the laboratory, the synthesis of amino acids may be carried out by the Strecker method, using hydrogen cyanide HCN in the presence of ammonia NH_3, and water H_2O:

$$R\text{-}CO\text{-}H \rightarrow R\text{-}CH(NH_2)\text{-}CN \rightarrow R\text{-}CH(NH_2)\text{-}CO\text{-}NH_2$$
$$\rightarrow R\text{-}CH(NH_2)\text{-}COOH$$

Taking the simplest case for example, starting with formaldehyde HCHO, we obtain glycine $NH_2\text{-}CH_2COOH$.

A trail leading to life

According to Cronin, 'with a bit of effort, it is possible to make amino acids from interstellar molecules, and the interstellar precursors are those required to make the organic compounds found in meteorites.' That being the case, we can draw up a table showing the interstellar precursors, the compounds found in meteorites (the building blocks of life), and the biological polymers that are the basis of life (see next page).

Micrometeorites

Although the table summarizes the reactions that may be obtained in the laboratory, is there any chance that they can occur in meteorites? A new aspect to this question has been raised by the study, not of meteorites, but of micrometeorites. Until 1984, these could

Precursors	Building blocks	Found in meteorites	Biological polymers
RCHO, HCN, NH$_3$, H$_2$O	amino acids	yes	proteins
HCN, H$_2$O	purines	yes	nucleic acids
HCN, H$_2$O, CHCCN	pyrimidines	yes	
H$_2$CO	riboses	no	
PN, CP ?	phosphates	yes	membranes
PAH* ?, polyynes ?	fatty acids	yes	

* PAH: polycyclic aromatic hydrocarbons (see p. 25)

only be collected in the stratosphere, or from marine sediments using magnetic separation techniques. Since then, however, Michel Maurette, CNRS Research Director at the Centre de spectrométrie nucléaire et de masse [Center for Nuclear and Mass Spectrometry] at the University of Paris at Orsay, has been able to obtain them from polar ice, initially from mud at the bottom of melt-water lakes in Greenland, and then from Antarctica, where they accumulate under the best possible conditions for their preservation.

In one of his recent collecting trips near the French Antarctic station of Durmont d'Urville, for example, by melting 100 tonnes of ice, he collected 10 g of sediments (no more!), containing 5000 intact micrometeorites, with diameters ranging from 50 to 200 microns. These had been able to survive passage through our atmosphere without damage. In his own words, it was 'the purest source of extraterrestrial micrometeorites ever found on Earth, containing chondritic grains in good condition.'

Working under a microscope, he patiently built up an itemized list of 300 grains. They were extremely porous, each resulting from the aggregation of small individual elements, less than a micron in size, consisting of silicates, oxides, and metallic sulfides. In particular, there were compounds similar to those found in carbonaceous chondrites such as the Murchison meteorite, with some that were even richer in carbon. This is the reason for the immense interest in this collection of tiny objects. Maurette has proved that these objects are of extraterrestrial origin, and that they have neither been altered chemically, nor contaminated biologically. It is, in fact, the organic component that acts as a protective heat-shield when these micrometeorites enter the Earth's atmosphere. It functions exactly like the ablative materials used by the aerospace industry.

Prebiotic seeding

Micrometeorites of the size studied by Maurette form the largest fraction, by mass, of the material collected by the Earth. This amounts to 20 000 tonnes a year, as against 100 tonnes (at the most) for all meteorites over 5 cm in size. This gives a new perspective on how the Earth is strewn with a massive amount of extraterrestrial organic compounds. The more so, because four billion years ago, toward the end of the major bombardment era, when our world had already cooled, the deposition rate may have been 1000 times as great. Our primitive atmosphere, which was possibly denser and thicker than nowadays, may have helped to provide a soft landing.

Let us imagine the scene suggested by Maurette: over 1000 years, each square meter of soil would collect a million micrometeorites. Each one of these fell on a different environment, perhaps even into a tiny, favorable puddle. Each grain formed a micro-environment, one-millionth of a cubic centimeter in volume, that contained various, highly concentrated ingredients and catalysts. The brief thermal pulse produced when it entered the atmosphere may have produced reactive chemical species. This microscopic environment had an enormous specific surface, because of the microscopic pores, vesicles, and cavities, all the way down to the molecular level. If it happened to fall on a damp, warm area of regolith, it became a miniature laboratory potentially capable of developing a prebiotic chemistry, which, if the views expressed by Cronin are correct, might have led to the synthesis of the building blocks of life from extraterrestrial material.

4

The primitive biological stage

We have now come to the start of bioastronomy's fourth stage, the one in which life finally appeared.

FROM AN INERT COSMOS
TO LIVING MATTER

At the very beginning of this book, I described the new paradigm that governs our view of the cosmos: we have progressed from a purely physical world to a biological universe. Biochemistry permeates the whole of space, ranging from interstellar material with its organic molecules to the heart of cometary nuclei. It has also existed ever since the Big Bang, with the exception of the first few hundred million years that it took for stars to be formed with carbon in their interiors and planets in orbit around them.

A new and crucial stage was reached when life emerged from inert matter. Where did this take place? Certainly it occurred on Earth, but then we do not yet know enough about the universe to say any more. When? It was a long time ago, about four billion years in the past. This enormous span of time corresponds to 100 million generations, from father to son, mother to daughter. If one member of each generation were to stand in rank and file, they would cover an area 10 kilometers square. Piled up as corpses, they would form a cube of side 400 m. This very distant time is indicated quite precisely by two events: first, the end of the intense bombardment that put the final touches to the formation of the Earth, and, second,

59

the appearance of the first fossils derived from living organisms: two facts, one cosmic in nature, and the other biological.

The lunar samples

Frequently, we hear people say that the exploration of the Moon added little scientific knowledge, and that the few hundred kilograms of rocks brought back by the Apollo astronauts have been sitting in inaccessible safe-deposits for the last 25 years. This is untrue, because thanks to those samples we have been able to reconstruct the intensity and chronology of the primordial bombardment. The Moon was also subjected to this bombardment; but its impact craters have not been affected by the geologic and atmospheric degradation that has prevailed on Earth. Study of its craters has enabled us to tackle the problem. A crater that breaks into another, for example, must obviously have been formed later; and a large crater corresponds to the impact of a body of significant size. Stratigraphy of lunar craters, linked with laboratory experiments, has enabled us to establish a relative chronology.

But we lacked calibration dates, and it is precisely these that we have been able to establish from the rocks brought back from the Moon. The 12 Apollo astronauts landed at six, different, carefully selected sites, and their extravehicular activities, initially on foot, but later using the lunar rovers, had well-defined aims. Their task was to pick up, by hand, *in situ*, and using informed judgment, samples that were later to be analyzed, after their return, in the laboratory. The rocks' radioactive isotopes have enabled us to calculate their precise ages, just as has been done for terrestrial rocks. Naturally, the rate of bombardment to which the Moon was subjected needs to be adjusted to apply to the Earth, because our planet is more massive and thus, throughout its history, has attracted more bodies than the Moon, even though, statistically, the number and size-distribution of meteoroids in interplanetary space was the same for both bodies.

The initial bombardment

After a simple gravitational calculation, we can obtain the curve showing the number of objects between 1 m and approximately

30 km in size that have fallen on Earth over the period of time between 4.5 billion years and 100 million years ago. This curve falls into two parts: one between 4.5 and 3.8 billion years ago, when there was a rapid, continuous, exponential decrease (by a factor of 100), and a much slower decrease (by a factor of 10), from around 3 billion years ago until the present. The transition between these two regimes corresponds to the end of the initial bombardment, which brought the formation of the planets to a close, and the beginning of the more general process that is still sweeping interplanetary space clean of debris.

Very recently, this curve, linked with the size distribution of meteorites, has served as a basis for more exhaustive calculations of the decline in the bombardment. This research is of particular significance for the question with which we are concerned: when were conditions calm enough for life to prosper in peace?

We find that 4.25 billion years ago there were still repeated falls of objects 500 km in diameter, any one of which was capable of completely vaporizing the oceans, not to mention causing deleterious effects on the atmosphere. We may recall that the extinction of the dinosaurs has been attributed to the impact of a body 10 km in diameter, 65 million years ago. Until about 3.8 billion years ago, however, the Earth was still subject to bombardment by objects 100 km in diameter, each one of which was capable of vaporizing the whole of the photic layer of the oceans, i.e., the top 200 m, precisely where there was some chance that life could appear. It is estimated that between 3.9 and 3.8 billion years ago, such impacts would occur every 10–20 million years.

It seems that because of this eventful early period, life would not be able to establish itself properly before 3.8 billion years ago. It might have appeared several times previously, and been subjected to extinction each time. According to Norman Pace, from the University of Indiana, certain life forms could have diffused down to the ocean depths, near submarine vents, where, thanks to the energy that the vents provide, they could have survived, protected by the seas above, and subsequently spread back toward the surface, when times were more propitious. In short, it appears that the crucial event, the transition from inanimate matter to life – although it did not necessarily have a well-defined beginning – may have taken place as early as 3.9–3.8 billion years ago.

The stromatolites

And what would be the latest date? A limit is provided by the oldest identified fossils. This is where stromatolites become involved. These carbonate formations, which are produced by the chemical activity of single-celled organisms, are still being formed today. They are found in shallow waters on certain parts of the Australian coast, where they form colonies consisting of domes measuring some tens of centimeters in diameter, and which resemble cushions or pouffes with a more or less spongy texture. They are produced by the accumulation of thin concentric layers, that trap mineral grains between them. The microorganisms that are responsible are cyanobacteria, such as the blue-green algae, which are photoautotrophic prokaryotes – in other words they live by photosynthesis, and by utilizing surrounding inert materials. Thanks to the energy in sunlight, they transform mainly water (H_2O) and atmospheric carbon dioxide (CO_2) into molecules of oxygen (O_2) and carbohydrates, or glucosides, such as the sugars $(CHOH)_n$. These chains consist of n repetitions of the basic structure HCHO, formaldehyde, which has already been mentioned frequently. For $n = 5$, we have pentose sugars, especially ribose and deoxyribose, which are the bases of the backbones of RNA and DNA. (In passing we may mention that, in these riboses, the carbon chain is partially replaced by a pentagonal ring consisting of four carbon atoms and one oxygen atom.)

Fossil stromatolites have formed impressive carpets ever since the early Precambrian; in fact, the oldest dated with reasonable certainty are those found in the north-west of Australia, at North Pole, which, despite its name, happens to be one of the hottest places in the world! The stromatolites' age is 3.5 billion years. It has even been possible to reconstruct their environment, which was a coastal lagoon in a marine volcanic landscape, rich in sulfates, which perhaps arose through the oxidation of sulfur by the oxygen that these ancient ancestral life forms released through photosynthesis.

The first parting of the ways

Note that we are closing in on the transition from inert material to life: the limits have been narrowed to between 3.9–3.8 and 3.5 billion years ago. Even more interesting results have been obtained, thanks, in particular, to the work by Manfred Schidlowski, from the Max-Planck Institute for Chemistry at Mainz in Germany. For a long time people have suspected the existence of even older fossils, dating back as far as 3.8 billion years, which narrows the gap for the formation of life rather alarmingly. Unfortunately, these fossil traces are hotly disputed.

Schidlowski has also tackled the problem by using isotopes of carbon. The most abundant, natural, stable nucleus is ^{12}C, consisting of six protons and six neutrons; this is followed by ^{13}C, with another neutron, which is far less abundant. (The ^{12}C:^{13}C ratio is 90:1.) There are also traces of ^{14}C, which is radioactive, and produced by the action of cosmic rays. It is this last isotope that is used for archaeological dating.

Biological separation of isotopes

Because of kinetic effects, the metabolism of life forms on Earth favors the use of ^{12}C, the lightest isotope. In the first place, when autotrophic microorganisms use atmospheric CO_2 at their active sites, CO_2 that incorporates ^{12}C diffuses more readily toward the sites, because it is lighter. Second, in reactions involving the formation of carboxyl molecules – i.e., molecules of the form R-COOH – which the cell carries out with CO_2, ^{12}C is again the most favored form.

In fact, when we measure the proportion of ^{13}C relative to ^{12}C in present-day living systems, we find that the former is impoverished by some 2–3 %, relative to the ratio found in inorganic minerals, such as the carbonates of nonbiological origin that are found in marine sediments. This finding extends to marine sediments that are of biological origin, and which consist mainly of kerogen (which itself consists of complex heavy polymers from dead matter, and graphite derivatives). These are also impoverished by 2–3 %.

More than 10 000 measurements of the relative proportions of ^{13}C

and ^{12}C have been carried out on sediments of all ages from around the world. In analyzing the measurements, Schidlowski found that certain common facts applied to a whole range of sediments, from contemporary ones to those as much as 3.5 billion years old – but not, however, to those that were 3.8 billion years old. These facts were: (a) the proportions of mineral (inorganic) carbons contained in carbonates have not varied; (b) neither have they varied in the organic carbon compounds in kerogens; and, above all, (c) the proportion of organic carbon relative to mineral carbon has not varied either, and has remained around 20 %.

Biological saturation

The biological implications of these findings are enormous. Ever since a period 3.5 billion years ago, a time very soon after the appearance of life, 20 % of the carbon has been organic – of biological origin. This fact is explained by a principle that may be called biological saturation: life proliferates exponentially until it reaches the limits imposed by the resources at its disposal. It is thought that the limit to life's expansion is, in fact, the supply of phosphorus, which is of vital importance in the structure of RNA and DNA. What we have witnessed, therefore, since a period 3.5 billion years ago, is global biological saturation by a prolific microbial ecosystem. This agrees entirely with the gigantic fossil deposits of stromatolites that have been found around the world.

Currently, these cyanobacteria are capable of producing 10 g of oxygen per square meter per day, if they are provided with unlimited food and energy, and if they are unharmed. It is the most highly productive ecosystem in existence, and it prevailed over the whole of the Earth during the Precambrian. Unfortunately for the bacteria, they sounded their own death-knell when they had produced enough oxygen for multicellular creatures to appear. These would be based on a far more efficient metabolism capable of burning that oxygen. The vast living carpets formed by the stromatolites were ravaged by gastropods, and nowadays they are found in just a few havens of highly saline water, such as in the Dead Sea. But without those gastropods would we be here to ask such questions?

Did life arise in less than 100 million years?

Schidlowski did not stop there. To try to resolve the question of possibly even more ancient life forms, he extended his analyses to the oldest sediments known, those from Isua, in western Greenland, which are 3.76 billion years old. He discovered that the organic sediments are 1 % less impoverished than more recent ones, whereas mineral sediments are impoverished by 0.2 % relative to recent deposits. These small, opposing variations, have, according to him, been caused by the metamorphosis of the rocks at temperatures of more than 600 °C, which would have allowed the different proportions of ^{12}C and ^{13}C isotopes to partially move toward an equilibrium state. Accordingly, the real proportion of ^{13}C has not varied since a time some 3.76 billion years ago. In consequence, similar life forms, and in particular, photoautotrophic forms, must have existed ever since that era. We thus obtain our second limit: life already existed 3.8 billion years ago. So between 3.9–3.8 and 3.8 billion years ago, life appeared on Earth and began to evolve. The interval available is less than 100 million years. It is astonishingly short.

From inert matter to life*

Ever since Pasteur's day, we have known that life has a past: it is an episode in the history of organic molecules – in other words of molecules that consist of carbon, hydrogen, nitrogen, oxygen, sulfur, and phosphorus – that owes much to the presence of a miracle solvent, liquid water. At a specific time in the history of the Earth, about four billion years ago, certain organic molecules began to produce copies of themselves that contained errors. The molecules were almost identical, but that slight amount of error enabled evolution by mutation to occur. Chemists therefore

---

* This discussion is based on remarks made by André Brack, Research Director at the CNRS Laboratoire de biophysique moléculaire [Laboratory of Molecular Biophysics] at Orléans, and François Raulin, Professor of Chemistry at the University of Paris, Val de Marne. I have summarized the points that they made in reply to the questions that I put to them during a broadcast on the radio channel France Culture, entitled 'From an inert Cosmos to the living Earth.'

have the task of reconstructing this primitive copying mechanism. They are placed in a situation rather like that of a wine-taster who is asked to describe the vintage, and the contribution made by the climate or the soil, and similar factors. As far as the 'vintage' is concerned, we are able to give an answer, because we have the basic information and some points of reference. Stromatolites, which are colonies of bacteria, have been dated more or less exactly: they may be traced back to 3.5 billion years. As regards organic molecules, we are at a disadvantage when compared with a wine-taster; if the latter has difficulty remembering or a problem with his taste buds, he can turn to the grower and ask him the year, which is carefully recorded in the appropriate ledgers. A chemist is unable to do that, because all traces of the organic molecules have, in fact, been erased, mainly by life itself, and also by one of life's by-products, atmospheric oxygen. In the absence of any trace of these organic molecules, we therefore have to resort to models, and to the experiments that they suggest. The models are devised by scientists' often fertile imaginations, and they are very numerous.

The classic scenario was proposed more than 50 years ago, by the Soviet biochemist Oparin: he suggested that chemical evolution's wonderful cuisine – to continue to use a culinary metaphor – developed in a primitive soup, which was constantly replenished by organic ingredients formed in the atmosphere. Expanses of water should have been abundant on the surface of the primitive Earth, and they could have played this role. This soup also needed ingredients, and Oparin's model assumes that the primitive atmosphere, which was different from the modern atmosphere, could have permitted the abiotic synthesis – i.e., synthesis without the presence of life – of organic compounds based on carbon. So an organic atmospheric chemistry, plus liquid water, might suffice for successful prebiotic chemistry. Are these merely the ramblings of a theoretician? Certainly not, because the model was tested experimentally in 1952–3 by Stanley Miller. He subjected a gaseous mixture of methane, ammonia, hydrogen, and water vapor to electrical discharges. The latter are a plausible source of energy in an atmosphere – we only have to think of lightning to appreciate that. After several days, his flask contained several organic compounds, in particular amino acids, which are the basic building blocks of life. This type of experiment has been widely undertaken since then, and the mechanisms involved have been studied in detail. At the time, there was doubt as to whether it would really be so easy to create the building blocks of life from a mixture with such a simple composition. That proved to be a surprise, but the results have been verified, and the validity of this type of experiment has been fully confirmed. It has thus been possible to demonstrate that most of the building blocks of life may be formed under such conditions. These are the amino acids, and the purine and pyrimidine bases that are the basic units of the RNA and DNA nucleic acids. It has also been realized that certain small, extremely simple, compounds, such as hydrogen cyanide and formaldehyde – the latter perhaps better-known in solution as formalin – are all that are required to synthesize practically everything required to construct the basic building blocks of life.

Starting with a small number of very simple, yet highly reactive, organic molecules (such as hydrogen cyanide, which has three atoms, or formaldehyde, which has four) that have the property of reacting spontaneously as soon as they are mixed with water, it is possible to produce all the compounds required for life, provided there is an adequate supply of the miracle solvent, water.

Nevertheless, building blocks do not make a wall. Before a truly living system could exist, all this had to be turned into a copying mechanism capable of amplifying information and transferring it to other systems. Once again, we had to begin with a model, because there are no fossil traces of any such rudimentary copying mechanisms. Initially, because we drew an analogy with present-day systems, we made the silly mistake of attempting to envisage a form of cell. It was, in fact, believed that any rudimentary or primitive copying mechanism among early life forms would resemble a cell, in that it would need a membrane to isolate the system from the aqueous medium surrounding it, and prevent the molecules and the information from dispersing into the sea. Quite apart from the molecules that formed the membrane, it would also be necessary to retain the molecules required to carry out the cell's essential chemical functions, work that is currently carried out by enzymes, a subdivision of proteins. Finally there would have to be yet a third category of molecules, the information molecules, capable of storing information and of transferring it to daughter molecules, functions that are carried out today by the nucleic acids RNA and DNA. It soon became apparent that the molecules with which we are most familiar are extremely complicated, and their appearance on the primitive Earth is highly unlikely. Consequently, it was necessary to decrease the degree of complexity required, and have recourse to far simpler molecules.

As far as the membranes are concerned, the new models are not very good. Although certain fatty acids may form vesicles, their synthesis requires temperatures of between 450 and 500 °C, a situation that is not plausible on the primordial Earth. As regards the catalytic molecules, the ancestors (at least to a certain extent) of the enzymes, the new models are generally good. People have succeeded in developing simplified models of enzymes, some properties of which have been reproduced using much shorter chains. These enzymes have only 10 links, and may be produced with a restricted number of amino acids. Instead of using the 20 amino acids that are found in current proteins, two different ones are adequate. Using 10 amino acids it is possible to carry out the work of proteins that contain 200. Such a reduction in numbers – from 200 to 10 – is extremely interesting. When it comes to work on molecules that can store information, the news is not so good. The molecules are very complex, and chemists, after 25 or 30 years of effort, have come to the conclusion that their synthesis is impossible under the simple conditions available in the laboratory. It would seem that this chemistry is too complex to have taken place on the early Earth. Life would not have started as a cell, and nucleic acids would not have been part of the first copying mechanism. The transition from organic molecules to living forms and the initial steps in establishing a copying mechanism probably took place without the assistance of nucleic acids.

Miller's experiments appeared to be on the right track, but subsequently people had to tinker with the composition of the primordial atmosphere. Miller chose a mixture of methane, ammonia, and hydrogen, because at that time it was thought that the primordial atmosphere had a composition similar to that of the giant planets, particularly Jupiter and Saturn. Since then, we have realized that there was never very much ammonia in the Earth's early atmosphere, because ammonia is a molecule that is readily broken down by the Sun's ultraviolet radiation. In addition, hydrogen is a very light molecule, which would have been able to escape easily from a planet as small as the Earth. Two of the components chosen by Miller had, therefore, to be excluded. All that is left is methane. It should be said that we have no direct indication of the composition of the Earth's primitive atmosphere; all that we can do is construct various models and see if they evolve into our current atmosphere. That does not work for ammonia and hydrogen. Similarly, the majority of geochemists do not think that the primordial atmosphere ever contained much methane. In addition, the Earth may be considered as having two sister planets: Venus and Mars. In both of these carbon is predominant in the form of carbon dioxide. The Earth's primordial atmosphere may, therefore, have consisted of carbon dioxide, water vapor, and nitrogen. Yet certain laboratory experiments have already established the organic compounds that are obtained and that depend on the initial gaseous composition. To create the basic building blocks of life, the mixture must include methane, and not carbon dioxide. This is extremely annoying for chemists studying prebiotic chemistry; the model by Oparin and Miller is just not suitable.

So we need to search for another prebiotic niche where the requisite elementary reactions could have taken place and created the necessary compounds. The hot submarine vents, discovered about ten years ago, are one possibility. Around them, we find gases that are discharged at extremely high temperatures, and also moderately reducing mineral salts that could have encouraged prebiotic chemistry. This is an interesting idea that is worth exploring, but, so far, it remains a theory, and there are practically no experimental models to confirm it.

We could also look to space. In the last 20 years or so, using radio astronomy, we have discovered more than 50 organic molecules in the interstellar medium, the largest of which consists of 13 atoms. As far as biochemists are concerned, this is too few. We can perhaps draw some encouragement from the fact that hydrogen cyanide, with just three atoms, has played an important part in the origin of life. Undoubtedly, more and more sophisticated, and more and more complex, molecules will be detected in space.

Amino acids are, however, very difficult to detect, because they do not have appropriate properties: they are elusive and, at present, none of the building blocks of proteins have been found in interstellar space. In addition, it is not easy to see how organic molecules in interstellar space could have traveled such great distances and ended up on Earth. The exploration of Comet Halley, however, has taught us that comets are packed with organic molecules. The comet's nucleus is dark, which certainly indicates that it is covered in hydrocarbons and organic compounds. The gases and dust have also been analyzed by mass spectrometers carried on board the probes. Cometary grains are far richer in organic matter than anyone could have predicted from observations from Earth. To judge by the impact craters on the Moon, early in its history the Earth was subjected to an intense bombardment by large meteorites and comets. Even closer to home there are the meteorites that you can pick up in your garden if you happen to be lucky enough to see one fall. Most of these do not contain organic matter, but some of them harbor up to 5% carbon, and detailed analysis of these meteorites reveals the presence of practically all the organic molecules you could dream of: several hundred are known so far. Ninety different amino acids have been identified (of which eight occur in our proteins), together with a whole range of analogous compounds. Micrometeorites collected in Antarctica form a new source of material, which will be subjected to analysis, particularly with regard to the organic compounds that it contains.

The case of Mars is interesting, because it now appears that early in its history, the planet was covered in liquid water. If it did have liquid water on its surface, and experienced the same rain of organic molecules as the Earth, then there is every reason for thinking that life, in a rudimentary form, would have developed there. In addition, the path life took on Mars could have been very different from what happened on Earth. This is yet another reason for going to Mars. As for Titan, it is the only satellite in the Solar System that has an atmosphere, quite apart from the fact that the surface pressure is 1.5 times that found on our own planet. Titan's atmosphere consists of nitrogen and methane. So it is not surprising that we have detected hydrogen cyanide, the basis of prebiotic chemistry on Earth, in particular, as well as many other organic compounds. Even though it lacks liquid water, and the temperature is too low, we can expect to find, if not life, at least some highly evolved prebiotic chemistry. Is this just a utopian dream? Certainly not, because the Cassini–Huygens mission currently being prepared will be launched by NASA and ESA in 1997. It will send a probe into the atmosphere of Titan and an orbiter around Saturn and Titan. By 2004, we shall have first-hand information about the chemistry. For us, Titan is an authentic laboratory, but one on a planetary scale. It offers us a fundamental environment where we can attempt to understand what happened when life appeared at a remote era in the Earth's early history.

The four acts in the drama

The appearance of life passed through four successive stages, each with its characteristic features.

> **Act 1** Simple organic molecules (which were synthesized in the atmosphere, in space, or at submarine vents, using water as a solvent) developed a prebiotic chemistry.

> **Act 2** An early copying mechanism, prone to errors that actually enabled it to evolve through mutations, became established, assisted by clays or protein sheets; this was the pre-RNA world.

> **Act 3** A simple ribonucleic acid that acted as its own enzyme, the 'ribozyme', combining the functions of information storage and catalysis, became established and evolved new functions, such as that of creating membranes and thus protocellules; this was the RNA world.

Act 4 The first microorganisms appeared, including our suggested ancestor, the 'progenote', which diversified and, about 2.3 billion years ago, gave rise to the three basic forms: the archaeobacteria, the eubacteria, and the eukaryotes. Much later, two branches of the eubacteria, the mitochondria and the chloroplasts, became symbiotic with the eukaryotes, eventually leading, about 700 million years ago, to plants, animals, fungi, protozoans, archaeozoans ... and the current world!

Which of these acts really saw the appearance of life? Probably either the second or the third. André Brack inclines toward the second, and suggests that people should search for a molecule that is self-reproducing. Several attempts have, in fact, been made to decide this question.

The crux of the matter

Given these problems, Gérard Spach, CNRS Research Director at the University of Rouen, has suggested that the sugar and phosphate backbone of nucleotides could be replaced by simpler chains, where ribose, a closed pentagonal ring, is replaced by glycerol, with an open ring. Yves and Éliane Merle, at the same university, have synthesized glycerides in this way, from which they have obtained short oligomers through dehydration on clays. They were able to initiate the polymerization of nucleotides, using an enzyme that would not have required a condensation matrix.

Marie-Christine Maurel, senior lecturer at the Pierre and Marie Curie University in Paris, does not replace the ribose with glycerol, but instead is investigating the pathway that begins with a modified adenosine (adenine + ribose), obtained by prebiotic synthesis. In this the ribose, instead of being fixed to the nitrogen atom known as nitrogen 9, in the hexagonal ring of the base (as is usual in normal biological systems), is fixed to nitrogen 6, the preferred form when adenine undergoes a condensation reaction with ribose. She has shown that this 'N^6-ribosyladenine' nucleotide fragment acts as a true catalyst, albeit a slow one.

This line of inquiry appears to suggest that the proteins and nucleic acids are not two distinct classes, one specializing in catalytic functions and the other in information storage. The Nobel prize for chemistry in 1989 was awarded to Thomas Cech and Sidney Altman, for their discovery of the catalytic properties of RNA. The question of whether DNA or proteins were the first to emerge has lost its meaning: they were preceded by a primitive form of RNA. The problem (which repeatedly arises in dealing with the origin of life) of knowing which came first, the chicken or the egg, is by-passed. More and more sophisticated experiments and the study of possible life forms on Mars will eventually allow us to understand how this came about.

PRIMITIVE BIOLOGY ON MARS

The martian tripods

No one would have believed, in the last years of the nineteenth century, that human affairs were being watched keenly and closely by intelligences greater than man's and yet as mortal as his own; ... During the opposition of 1894 a great light was seen on the illuminated part of the disc, first at the Lick Observatory, then by Perrotin of Nice, ... Then came the night of the first falling star. It was seen early in the morning rushing over Winchester eastward, a line of flame, high in the atmosphere ... Ogilvy, who had seen the shooting star, and who was persuaded that a meteorite lay somewhere on the common ... rose early with the idea of finding it. Find it he did, ... An enormous hole had been made by the impact of the projectile, ... The uncovered part had the appearance of a huge cylinder, caked over, ... Then suddenly he noticed with a start that some of the grey clinker, the ashy incrustation that covered the meteorite, was falling off the circular edge of the end ... And then he perceived that, very slowly, the circular top of the cylinder was rotating on its body ... The cylinder was artificial – hollow – with an end that screwed out! ... At once, with a quick mental leap, he linked the thing with the flash upon Mars.

The scene is set with the arrival of the martian lander on Earth. Then two nights later:

The thunder-claps, treading one on the heels of another and with a strange crackling accompaniment, sounded more like the working of a gigantic electric machine than the usual detonating reverberations ... Then abruptly my attention was arrested by something moving rapidly [toward me]. At first I took it for the wet roof of a house, but one flash following another showed it to be in swift rolling movement ... And this thing I saw! How can I describe it? A monstrous tripod, higher than many houses, striding over the young pine-trees, and smashing them aside in its career; a walking engine of glittering metal, striding now across the heather; articulate ropes of steel dangling from it, and the clattering tumult of its passage mingling with the riot of the thunder. A flash, and it came out vividly, heeling over one way with two feet in the air, ... Can you imagine a milking stool tilted and bowled violently along the ground? ... Seen nearer, the thing was incredibly strange, for it was no mere insensate machine driving on its way. Machine it was, with a ringing metallic pace, and long flexible glittering tentacles (one of which gripped a young pine-tree) swinging and rattling about its strange body. It picked its road as it went striding along, and the brazen hood that surmounted it moved to and fro with the inevitable suggestion of a head looking about it ... As it passed it set up an exultant deafening howl that drowned the thunder, 'Aloo! aloo!' ... I began to ask myself what they could be. Were they intelligent mechanisms? Such a thing I felt was impossible. Or did a martian sit within each, ruling, directing, using, much as a man's brain sits and rules in his body? I began to compare the things to human machines, to ask myself for the first time in my life how an ironclad or a steam-engine would seem to an intelligent lower animal.

Here we have a martian tripod, as described by H. G. Wells in one of the first, and best, science fiction novels, *The War of the Worlds*, published in 1898, nearly a century ago.

The first landings

By an irony of fate, or a happy turn of events for Earthlings, it was they, and not the Martians, who deployed tripods on the neighboring world! On 20 July and 3 September 1976, Vikings 1 and 2 gently landed on the desert plains of Chryse and Utopia. This high point of technology, achieved by NASA, revealed to humans the first panoramas of Mars, and crowned ten years of effort by the United States and the Soviet Union. After the exploratory fly-bys by the American Mariner space probes in 1965 and 1969, the Soviets soft-landed two capsules on the surface in 1971. Unfortunately these did not send back any signals, and there were further set-backs for the Soviets in 1973: one capsule was lost in space, and the other did not send back any data. Among these Soviet attempts, we now learn, there was even a rover. The conquest of Mars has proved to be difficult, because later probes, the two Soviet Phobos missions in 1989, proved to be semi-failures, or semi-successes. The same thing happened to the American Mars Orbiter mission in 1993! Another attempt was to have been made in 1994, but has been postponed for several years for economic reasons ... When shall we see human beings on Mars?

The first habitable world

Mars has always exerted a powerful attraction for people on Earth. Every two years it disappears, to stage a spectacular reappearance some time later as a fiery red object in the sky. Its blood-red color has inspired terror and war. It was only after the theoretical work of Copernicus at the end of the 15th century, and the observations of Galileo at the beginning of the 17th, that Mars became just another world. Copernicus showed, in fact, that this red 'star' revolved around the Sun, like our own Earth, in what subsequently became known as a planetary orbit. Galileo, observing the heavens with one of the first telescopes, saw that Mars was a globe, similar to our own. This is when it became a planet in its own right. As a lover of old astronomical books – although limited by their astronomical price – I have the French translation of what is described as the 'New Treatise on the Plurality of Worlds by the late M. Huygens, a member of the Royal Academy of Sciences.' This 1702 edition carries a preface by Monsieur de Fontenelle, of the Académie Française, who wrote a eulogy of the work: 'By order of Monseigneur the Chancelier, I have read the present manuscript, and I believe that

the public will not fail to derive pleasure and utility from the translation of the last work by such a great man as the late M. Huygens.'
In 1686, Fontenelle himself published his famous book *Entretiens
sur la pluralité des mondes* ['Discussions on the Plurality of Worlds'],
and it is pleasing to see a great philosopher recognizing the merits
of a great astronomer in such plain language.

Meudon and extraterrestrial life

This attraction for Mars, and the question of whether it was inhabited, was strengthened when large refractors were introduced in the
19th century. The one at Meudon (which, with an objective 83 cm
in diameter, is the fourth largest in the world) is a perfect example.
On the occasion of the public session of the five French Academies
on 24 October 1896, the Director of Meudon Observatory, Jules
Janssen, spoke on the subject of 'Extraterrestrial life, and the existence, outside the Earth, of worlds more or less similar to our own.'
He said:

> ... Any scientific study of extraterrestrial life must begin with
> the study of the planets ... Their disks show indications of conti-
> nents, clouds, and atmospheres ... These similarities in physical
> constitution are palpable, established facts ... The unexpected
> discovery of a new method of investigation has enabled us to take
> a new, and decisive step, concerning this question. We mean
> the discovery of spectral analysis ... During a visit to Etna, un-
> dertaken in an attempt to avoid the deleterious effects of the
> atmosphere, a French physicist [in fact, himself] confirmed the
> presence of water vapor in the atmosphere of Mars ... This sim-
> ilarity bears witness to an even more general similarity than just
> the overall physical constitution of these bodies.

This is how Janssen, a pioneer in the spectroscopy of planetary
atmospheres, broached the idea that water, the fundamental basis
of life, might exist on Mars. In the first volume of the *Annals of
the Observatory of Meudon*, Janssen, referred to 'the possibility of
approaching the study of the chemical composition of planetary atmospheres, and through that, of taking a new, decisive step regard-
ing the question of the habitability of the worlds and extraterrestrial

life', and wrote: 'France must not fail to explore this avenue that we have begun to investigate in such a propitious manner ... The authorities have understood this and, anxious to uphold the honour of French Science, have decided to create an observatory specially devoted to physical Astronomy.' One could say, therefore, that it was because of extraterrestrials that one of the greatest observatories in the world was born, a hundred years ago, on the slopes of Meudon. Nowadays 500 people work at the observatory.

Earth's sister planet?

The intense interest in Mars had already been encouraged by observations made in the 19th century, with far smaller refractors than the gigantic one at Meudon. Mars appeared to be a globe half the size of the Earth, with a year twice the length of our own, days of nearly 24 hours, an inclination almost identical to that of our own world, and therefore with four seasons resembling our own. It had two polar caps that increased and decreased in size with the seasons, continents that resembled ochre-colored deserts, and even martian 'seas' of a blue-green color that lightened or darkened according to the melting of the polar caps.

A dream world! To such an extent indeed that even nowadays, because it lies within the scope of modest-sized telescopes Mars is the favorite object for amateur astronomers. A good pair of binoculars, magnifying eight times, is enough to give one the pleasure of seeing craters on the Moon. Even two poor lenses will suffice: hold up a magnifying glass and the lens from a pair of spectacles that correct for far-sightedness, adjust the focus and you can glimpse lunar craters, albeit not very sharply, and washed-out in appearance. For Mars, a good-quality refractor with an aperture of at least 10 cm is required, or a reflector, perhaps like a Celestron 8, accurately aligned, and set up on a good mounting, in a good site free from pollution and atmospheric turbulence. Above all, one needs experience in observing, and perseverance, because, even with a magnification of 200 times, the disk of Mars is very tiny when seen in the eyepiece. In addition, it is delicately shaded: the ochre and blue-green tints show little contrast, and the polar caps are very small. But observation can be rewarding. Even a professional astronomer can find pleasure in 'seeing' Mars under these conditions, despite the infinitely superior photographs that have been sent back by space probes.

During the 1986 opposition, I was able to use a good-quality 10-cm refractor. I watched the way the southern polar cap melted and how the northern cap become covered in a layer of mist. I also followed a dust storm that was born in the depths of the giant Hellas impact basin (7000 m deep!) and, for eight days, observed it moving toward the west and developing until it covered twice the area of France ...

Traces of water

The quest for water on Mars has motivated many astronomers. Audouin Dollfus, a well-known astronomer from the Paris Observatory at Meudon, continued the tradition. He had enough courage to allow himself to be carried up into the stratosphere, in an air-tight sphere that he had built, beneath a cluster of 100 meteorological balloons. His aim, like that of Janssen, was to escape from the water vapor in our atmosphere so that he could attempt to detect water vapor on Mars. I take my hat off to him, because when I worked with Louis Leprince-Ringuet, I used small clusters of half-a-dozen such balloons to lift a few kilograms of instruments intended to observe cosmic rays. To the detriment of the instruments, but not, luckily, of me, I was frequently reminded of their fragility!

The space probes have shown definitively that the amount of water vapor in the martian atmosphere is only 0.03 %. If it were to be condensed, it would form a layer on the surface with a thickness of just one-tenth of a millimeter: not a very deep ocean! This has not prevented water on Mars from inspiring some fantastic tales. Some magnificent landscapes, in particular, were drawn by the amateur astronomer, painter, and scientific writer, Lucien Rudeaux. Long before the conquest of space, this world pioneer of astronomical painting had created representations of scenes that have decided many people in their choice of vocation, and promoted this world of the future. One such example is the fly-by of Phobos as seen from a spacecraft approaching Mars, which may be found in his aptly named, 1937 masterpiece, *Sur les Autres Mondes* ['On Other Worlds'].

The Viking Landers' biological experiments

The hypothesis that there was life on Mars even became the justification for sending the Viking 'tripods.' Each lander carried a tri-partite biochemical laboratory, designed to detect the metabolic processes carried out by primitive biological life forms. A small grab scooped up from the surface samples of the fine dust that had been deposited by the winds, and dropped them into the laboratory's interior. These were the first earthworks carried out by our civilization on another planet! The samples were provided with various nutrients, some of which were radioactive and some not, and then incubated; the by-products were then analyzed by pyrolosis, gas exchange, and by the degassing of chemical tracers. The instruments, which were of a complexity never previously attained, had to be accommodated in a volume of less than 30 litres. Despite the innumerable possible situations that had been simulated in advance, as well as those carried out under more faithful martian conditions, the interpretation of the results was extremely difficult. The surface regolith contained oxidizing compounds but no organic compounds. Consequently, no organic life forms are to be found in the martian deserts. Nevertheless, the absence of organic compounds poses a problem, because the continual meteoritic bombardment, even if limited in amount, should deposit some organic material on the surface. Presumably, destructive surface processes have erased any traces of life. It is not, therefore, possible to state that life never existed on Mars. Should we seek elsewhere than in the desert plains chosen by the cautious NASA engineers? Do we need to explore more varied regions, such as the polar regions, or deeper layers of the soil?

A desolate landscape

The landscape around the Viking Landers is cold, desolate, and dry. The atmosphere is thin, windy, injurious, and asphyxiating. The meteorological stations on the Viking Landers provided daily observations for two complete martian years (four Earth years). They showed temperatures varying from −30 to −90 °C, depending on the season and the time of day, and pressures of 700 and 1000 Pa (7–10 millibars), very similar to the environment around an airplane in mid-flight. In calm weather, the wind-speed is between 10 and 25 km/hour, but during storms it rises to 70 km/hour, and even 400 km/hour in dust storms, which raise thick curtains of dust that hide the Sun for weeks. Sometimes it takes months for the atmosphere to clear.

The atmospheric composition is 95 % carbon dioxide, 3 % nitrogen, and only 0.1 % oxygen: utterly unsuitable for any animal life. The soil is rocky, covered with slightly compacted sand, across which, driven by the wind, dust and small pebbles sometimes patter like hail. This only remotely resembles the conditions found in deserts like those on the high plateaux in Chile, around the European Southern Observatory, or the frozen wastes of the Ross dry valleys in Antarctica.

Water on Mars!

Despite their disappointment, astronomers have not been discouraged. There was considerable surprise when the Viking Orbiters that had accompanied the landers to Mars revealed other martian landscapes. These Orbiters took 51 000 photographs of the surface, some of which had a resolution of as little as 10 m. Imagine Mars being watched by a pair of powerful eyes: the smallest church, and the smallest ship would be visible. Again, we must pay tribute to American technology at the time; this happened 15 years ago, and the spacecrafts were controlled and the data transmitted over a distance of some 300 million km. The photographs, distributed in the form of magnetic tapes to the principal research institutes around the world, have enabled us to study the whole of the martian globe. Its general appearance is that of a inanimate, cold, desert world. But there are some notable features, including volcanoes, one of which, Olympus Mons, with a base 700 km across and a height of 27 000 m, is the largest in the whole Solar System. A gigantic canyon system, Valles Marineris, forms an immense gash in the crust, 9000 km long,

100 km wide, and up to 6000 m deep. There are enormous impact basins, such as Argyre, which is 600 km in diameter and 1000 m deep.

Above all, however, the great surprise was the dried-up river-beds, some as much as 15 km wide, and whose discharge must have been 1000 times that of the Amazon, our greatest river! If liquid water existed on Mars in such quantity, would there not have been a denser atmosphere, which would imply a far more temperate climate, perhaps favorable to the spontaneous appearance of life?

Kasei Vallis

The finest dry river valley is Kasei Vallis. Its bed contains islands that have been molded by the stream into elongated tear-drop shapes. Where the stream turned a 90° bend, the water spilled over onto the neighboring plain, where it left characteristic gullies. The river was born suddenly, broad and fully formed, at the end of an area of chaotic terrain, a vast tract of land, 40 km on a side, that has been created by collapse of immense blocks of the surface, between which there are mesas, undisturbed parts of the original plateau. It is believed that a vast underground reservoir of liquid or frozen water suddenly overflowed, released by nearby volcanic activity. (Similar instances are known on Earth, for example in the eastern region of the State of Washington, and in Montana, where the collapse of a glacial dam that retained Lake Missoula released a flood 120 m high, with a discharge rate 100 times that of the Amazon, and which lasted several days, leaving channels 200 m deep, caused by massive erosion.)

Beginning at this area of chaotic terrain, the flood swept down a vast, ill-defined depression, 300 km wide by 1500 km long. Downstream, the gorges of Kasei Vallis broaden out in the plains around Chryse, 9000 m lower, where they are joined by other streams. By counting the number of impact craters it is possible to determine that the age of these river systems is between 3 billion and 3.5 billion years, although there may have been much more recent episodes of flooding.

The channels

Other types of flooding are more classical in form: the stream is restricted to a distinct, well-defined, narrow, deep bed, with numerous tributaries. They correspond to what would be expected from the gradual drainage of an underground liquid or frozen reservoir. The channels must have gradually extended headward, with the water flowing gently over the surface. Channels of this type are found on ancient landscapes and it is these, rather than the catastrophic floods, that may indicate the existence of a temperate climate on Mars prior to three billion years ago. But not all astronomers are in agreement on this point.

We should look at the 51 000 photographs taken by the Viking Orbiters with this association with water in mind. Let us imagine for a moment that we are approaching Mars on board one of the Orbiters: flying over the jumbled landscape we can see the thin atmosphere outlined on the horizon, where a few thin layers of high-altitude clouds are visible. Next, we find ourselves flying over a cyclone, a magnificent spiral of cloud, as perfect as an illustration in a meteorological book. At the bottom of the canyons and depressions there are thin morning mists. In the distance, a vast, swirling mass of dust is being carried aloft by a storm, which casts a distinct shadow on the ground beneath. It looks just like the terrifying front of an approaching sandstorm over the Sahara desert or the plains of Burkina Faso. We cannot fail to be impressed by the vitality of the martian atmosphere: it would not take much for us to open the hatches and take a deep breath of fresh air!

The frozen subsoil

This link with water is brought dramatically to our attention in yet another fashion: thanks to the work of François Costard, of the Laboratoire de géographie physique [Physical Geography Laboratory] at Meudon-Bellevue, we have discovered the implications of the lobate impact craters. Imagine throwing a stone into a pool of almost dry mud: it would form a crater surrounded by ejecta, which would come to rest all round it but not spread out very far. On Mars, Costard has detected 2000 such craters. The lobes are generated when the frozen subsoil – a permafrost layer – is melted by the heat of the impact, and immediately re-frozen at the surface. The larger the diameter of the crater, the more powerful the impact, and the deeper the projectile penetrated into the ground. By measuring the volume of the permafrost ejected as a function of the depth, it is possible to calculate the depth of the layer of frozen subsoil. The top of the martian permafrost lies at about 300 m near the equator, and at around 100 m in middle latitudes. In places, such as in Chryse Planitia, where the formation of permafrost was assisted by the catastrophic flooding, it may be present at a depth of only 60 m. This is where human explorers should go to fetch their water. As for its thickness, it may be estimated from the geothermal gradient, which, on Mars, possibly corresponds to an increase of a few degrees Celsius for every 100 m of depth. That gives a permafrost layer several kilometers thick, twice that found in north-eastern Siberia. It is an enormous reservoir of water.

This permafrost consists of a mixture of fractured rocks (or regolith) and interstitial ice: it is typical of dry periglacial regions. Pockets or films of liquid water may exist within the permafrost, and these must become more frequent below 4 km in depth. In conclusion, if life ever existed on Mars, we need to look for it in the frozen subsoil.

The polar caps

To avoid being forced to dig very deep, it would be easier to choose the polar regions. This is because the polar caps are important agents in the water cycle. The Orbiters obtained images of them that showed extraordinary detail. The two caps are very different, because Mars, with its fairly elliptical orbit, is closer to the Sun during the southern summer.

The southern cap has a hot summer and a long cold winter, which means that it has a greater maximum extent and a smaller minimum area than the northern cap. At the end of winter, it extends as far north as latitude 50° and, at the end of summer, it has shrunk to some 350 km across. Its structure during the summer looks like a whirlpool because of the spiral nature of the terrain, where residual ice remains at the bottom of craters or on southern slopes. In 1969, the cap even disappeared completely. The residual northern cap is generally some 1000 km in diameter, and regularly has a spiral appearance where it lies along the sides of valleys, some of which are several hundred meters deep.

During the winter, the upper layers of the caps consist of dry ice (frozen carbon dioxide) which may be 50 cm thick. In summer the residual northern cap probably consists of water ice. But it is difficult to estimate the size of this reservoir.

An unstable climate

The southern polar cap retreats very rapidly. This behavior, together with the effects of the deep impact craters in the region, tends to favor the sudden onset of dust storms. This may partly explain the difference between the two caps, because the carbon dioxide in the atmosphere freezes onto the nuclei provided by the dust raised by the storms. This tends to transport the carbon dioxide from the south toward the north. The northern polar cap is dirty, which encourages the carbon-dioxide ice to melt completely during the summer.

These complex north/south interactions involving carbon dioxide, dust, winds, and orbital eccentricity reverse from time to time, because of an oscillation that occurs in the angle of the planet's axis of rotation. This is like that of a top that is slowing down, and it has a periodicity of 25 000 years. This is almost certainly the origin of the layered terrain found in the polar regions. On ice-free slopes of the northern valleys it is possible to distinguish thin strata, consisting of annual deposits of dust, each a few tens of meters thick. Their total thickness may amount to several kilometers. The deposits bear witness to climatic variations on Mars over the last few million years. They have been simulated on a computer – as have the Ice Ages on Earth – taking into account the changes in the eccentricity of the orbit and in the inclination of Mars' axis of rotation.

The results are astonishing, and we can count ourselves lucky that similar climatic changes have yet to happen to us. Every 1.2 million years, the pressure fluctuates between 100 and 4000 Pa (1 and 40 millibars), with, moreover, major oscillations every 100 000 years.

The carbon dioxide may freeze out completely at the poles, reducing the atmosphere to insignificance. To these effects we also need to add the greenhouse effect, the ways in which dust storms are triggered, and the absorption of gases by the regolith, all of which are yet to be fully understood. Although at present the north pole is accumulating debris and the thickness of layered terrain is increasing, this will change in a few thousand years, with consequences that are hard to predict.

In addition, it is possible that the crust is shifting above the mantle and relative to the axis of rotation. Much older layered deposits exist elsewhere closer to the equator. When we send our mobile robots to Mars, it may well be here, on the eroded slopes of these layered deposits, that we find fossils of primitive life forms.

According to Chris McKay, of NASA's Ames Research Center, other sites are also favorable: in particular the sedimentary deposits left behind by ancient lakes that existed before 3.8 billion years ago, and which have left thick sequences of carbonate rocks, especially in the Tharsis volcanic region, and in the various canyons forming the Valles Marineris system.

So we shall be able to search for life on Mars. Obviously it will be a primitive form of life, but a marvellous example is shown by terrestrial stromatolites (p. 21). These are concretions left behind by single-celled organisms that have persisted for thousands of millions of years: ten times as long as the period dominated by the dinosaurs, those masters of successful multicellular life, and a thousand times as long as the period that saw the development of *Homo sapiens*, another form of master, but this time of information processing, and thus of what we like to call intelligence.

Terrestrial equivalents

In searching for potential terrestrial equivalents, we can call on the help of numerous scientists working in various disciplines. For example, permafrost specialists, such as D. A. Glichinsky, from the Institute for Soil Science and Photosynthesis in the former Soviet Union, and E. Brock, from the Institute of General Botany in Hamburg, have found active nitrogenous bacteria in the frozen

Siberian subsoil at a depth of 35 m, where the layer is several million years old. These microbial ecosystems have a greatly reduced metabolic rate, and occur in the liquid water found in bubbles, pores, and thin films in the Arctic permafrost. Under these difficult conditions, the bacteria have survived, encysted, in a cryogenic system for three million years, and resume a normal rhythm when conditions improve. These researchers hope to extend these results back to 30 million years by investigating the Antarctic permafrost, which is much older. By means of extrapolation, they may then be able to estimate the chances of long-term preservation on Mars.

L. H. Hochstein, from NASA's Ames Research Center, has found hyperhalophilic bacteria in ancient evaporites, the remnants of saline environments that dried up in the Triassic and Permian. He thinks that the growth of such bacteria would be possible on Mars in the presence of water. To him, these organisms are possible models of martian life, and may, one day, play an important part in its exploration and simulation.

A. H. Segerer, who holds the chair of microbiology at Regensburg in Germany, is interested in hyperthermophilic microorganisms that live in volcanic habitats; the optimum temperature for these organisms is above 80 °C, and they can survive up to 110 °C. They are chemolithoautotrophs (gaining nourishment from the chemical breakdown of stone), whose metabolism is based on the oxidation of hydrogen and reduced sulfur compounds. They are thus completely independent of solar energy and could survive in the presence of water and volcanism, even outside the solar ecosphere. These organisms belong to both the bacteria and the archaeobacteria. Hyperthermophilic characteristics are thus very ancient, which is an argument in favor of life on Earth having arisen at sites similar to the submarine vents.

Another interesting terrestrial environment is that of the frozen lakes found in the cold, desert, dry valleys of Antarctica. Although the mean temperature of these deserts is −20 °C, the deep lakes contain liquid water beneath their permanently frozen surfaces. According to C. P. McKay and L. R. Doyle, of the SETI Institute in California, these lakes might serve as models of martian habitats where primitive life forms may have survived.

What sort of life should we seek on Mars?

According to the supporters of life on Mars, we should seek every possible form: microfossils, cellular material, altered organic compounds, microstructures similar to stromatolites, and biomineralizations. After all, two-thirds of the surface of Mars consists of ancient terrain, which is accessible, albeit heavily cratered, and the photographs obtained by the Viking Orbiters have enabled us to identify the sites of possible lacustrine sediments.

These ideas, strongly supporting the search for life on Mars, are very recent. I remember that during the international colloquium on the preliminary results of the Soviet Phobos probes, held in Paris by the Centre national d'études spatiales [National Center for Space Studies] (CNES) toward the end of 1989, V. I. Moroz, from the Institute for Space Research in Moscow, described projects for future missions, in particular the penetrators that would be dropped deep into the martian surface. Although each would weigh 50 kg, no instrumentation had been included for the detection of biological activity. Since then, interest in bioastronomy has gained ground in the Institute's planning.

Martian simulations

The German Air and Spaceflight Research Association (DLR) in Cologne has a space simulation chamber, which, since 1987, has been used to study the surface of comets under appropriate conditions. In this giant chamber, which is a cylinder 2.5 m in diameter and 5 m long, evacuated and cooled by liquid nitrogen to $-190\,°C$, samples of soil 80 cm in diameter are subjected to 65 kilowatts of radiation, equivalent to twice that from the Sun. This chamber allows life-size experiments to be carried out to clarify the specification for the Rosetta probe (p. 38), and even to estimate the behavior of its harpoon and coring tool. Currently, the DLR is about to begin simulating the martian soil with appropriate minerals, temperatures, and carbon dioxide atmosphere.

A. Banin, from the Weizmann Institute of Science at Rehovot in Israel, has provided a model of the soil. The Viking probes, in particular, have shown that the regolith on Mars has evolved through the aqueous oxidation of primitive basaltic rocks exposed to the atmosphere. This has broken them down and their components have been thoroughly mixed by global storms. Taking account of the observational constraints imposed by the color, reflectance, chemical composition – silica is the principal component – microscopic properties, and magnetization, the fine-grained material is well represented by a mixture of clays, such as montmorillonite with amorphous oxides of iron, such as Fe_2O_3. This is probably what will be used in the chamber at DLR. Investigations will determine the depths to which lethal solar ultraviolet radiation penetrates, the behavior of water ice (such as its condensation, diffusion, recrystallization, and reactions when bombarded by cosmic rays), and related problems.

Exobiological simulations

As regards exobiology, the simulations will be undertaken to assist in laying the foundations for various lines of research: the search for traces of life, past or present; methods of preventing the planet from being contaminated by the exploratory equipment itself; the creation of artificial ecosystems; and the study of similar terrestrial environments.

These are some of the projects envisaged by Gerda Horneck, the Director of the DLR's Institute of Flight Medicine, which has a long history of research in this field, in particular with the exposure of bacterial spores to space on board LDEF(Long Duration Exposure Facility). This satellite was intended to be brought back to the ground by the Space Shuttle after a flight lasting a year, but had to remain in space for six years following the *Challenger* disaster. This meant that the set of spores being used as a control in the DLR laboratory had to be maintained for six years, and kept under appropriate temperature and lighting conditions. It was found that 90 % of the spores of *Bacillus subtilis* survived under the extreme conditions found in space, when they were protected by a layer of glucose, or simply by other spores of the same kind.

Various characteristic states of the martian climate will be simulated to determine the limiting factors that would govern the presence of life, using varying conditions of atmospheric composition, pressure, temperature, exposure to sunlight, and porosity and density of the regolith. Much is also being learned from the experiments that were carried on the European Recoverable Carrier (Eureca), during the mission that exposed other biological material to the conditions in space.

The biological exploration of Mars

The biological exploration of Mars is based on the idea that life appeared on that planet four billion years ago. Subsequently, it either disappeared 3.8 billion years ago (which is why we need to search for fossils), or else adapted to current conditions (which is why we need to search for surviving ecological niches). Particular attention will be devoted to studying conditions around the northern polar cap, where some of the residual water sublimates during the summer. The behavior of organic material at the surface will be analyzed under laboratory conditions, with the aim of understanding the negative results from the Viking Landers. The growth of selected microorganisms that have been introduced into certain highly specialized niches, such as saline crystals, will help to clarify the processes of chemolithoautotrophy. These simulations will enable us to guide the vehicles toward the most promising areas, and will help to interpret their observations.

The martian rovers

Our knowledge of Mars will be greatly advanced by the use of rovers. The Viking Landers carried out an initial reconnaissance, but, being stationary, they were able to study only the few meters of surface in their immediate vicinity. For observations elsewhere, we expect to use a network of small fixed stations, such as penetrators. But more is required: we need to move about to find special samples – the local Rosetta Stone – which may be hidden behind a giant boulder, or staring the lander in the face from the opposite side of a crevasse.

Although the Apollo astronauts, complete with mobile phone, were able to drive around the Moon on their all-terrain vehicle, and thus search for unusual rocks and bring them back to our laboratories, it was the Soviets, who, using the first extraterrestrial rovers, the Lunakhod vehicles, carried out the task by remote control. These 700-kg devices, with their eight wheels and stereoscopic cameras, covered 10 km on the surface of the Moon, guided from a center situated in the USSR.

Marsokhod

It is hardly surprising that their successor comes from the same stable: after the failure of its predecessors in 1971, the latest Marsokhod should be launched on its journey to Mars within a few years. The extraordinary device rolls and walks at the same time: its six independent wheels can turn and lift. Each one of them is a combination of a cone and a cylinder, ensuring the maximum degree of contact with the soil, even if it is loose or angular in nature. The chassis is articulated, and capable of various contortions to follow the terrain, and is also able to extend or shorten itself, rather like a caterpillar. For example, it can anchor its four rear wheels, and, using its two front wheels, excavate the soil to get past an obstacle. It can then pull its rear wheels forward.

Marsokhod weighs just 100 kg; each wheel consumes only 4 watts of power; and it can travel at 500 m/hour. I have seen a film taken during trials among the rocks of Kamchatka. It was amazing! Like a metallic weevil, it crouched on the ground, hugging the bumps and hollows, resembling an insect with swollen feet. Seeing this vehicle, the Martians will be hard put to imagine any of our vehicles, whether horse-drawn gigs, Rolls-Royces, Boeing 747s, or high-speed trains!

The device carries coring equipment with a total weight of 4 kg; it can excavate 6 cm^3 of soil, 12 mm deep, five times per minute. In loose soil, it is able to bore down as far as 2 m. The samples are transferred to the analytical equipment, which includes a gas chromatograph and its pyrolyzer, together weighing less than 2 kg. These are able to analyze organic and volatile compounds. Marsokhod will, in effect, have a nose capable of sniffing out any possible biological activity!

If, in the course of its wanderings, it bumps up against a rockwall, or encounters the edge of a fault, the obstacle is recognized by sensors located on the forward part of the vehicle. It then backs, turns slightly, and moves forward again. If it succeeds this time, it resumes its previous course, or else tries again. This method of operation is costly in time and energy. However, we must not forget that, unlike Lunokhod, which was close by on the Moon, basic maneuvers cannot be controlled from Earth. In case of danger, it would take 20 minutes for it to warn us, and as long again for us to send it directions of what to do, because radio waves merely travel at the speed of light. This is why the CNES suggested to the Russian scientists that the vehicle should be fitted with stereoscopic vision and a brain: twin cameras and a computer that would enable the rover to build up a relief map of the terrain in front of it. It would then choose the best path by which it could accomplish its mission, in accordance both with the Earthlings' wishes and with the information that it had gathered on the spot.

The CNES rover

This initial scheme for a degree of artificial intelligence on the martian rover opens the door to possible future developments. Francis Rocard, the director of the CNES martian project, is the grandson of Yves Rocard, well known and highly regarded by French radio astronomers because he was largely responsible for the erection of the large radio telescope at Nançay. Francis Rocard and his team, together with other French scientists, are working on a project for a more complex rover. This would have a mass of 800 kg, and would be able to travel 1000 km over a period of more than two years. It would be given the task of establishing a base station on the surface (similar to those deployed on the Moon), with a seismometer, magnetometer, meteorological instruments, solar panels, and central laboratory. There would then be two other, smaller stations. The rover would then explore the surrounding area, obtaining geophysical profiles of the subsoil by radar sounding, and gravimetric and magnetic measurements. Thanks to two arms fitted with vision systems, sensors and various tools, its mobile laboratory will be able to carry out chemical, mineral and dating analyses.

Its artificial intelligence will allow it to choose the best movements, and to select, acquire, and then analyze suitable samples. As a test, it is proposed to simulate a trip over a distance of several hundred kilometers across the ancient flood plains around Kasei Vallis, with detours to sites of geophysical and geomorphological interest. Five instruments specially dedicated to exobiology, including the all-important chromatograph, may, perhaps, reveal the greatest discovery of the 21st century.

It is easy to understand how, for Francis Rocard, mobile robots represent one of the greatest challenges to technology, because they combine problems in the fields of both artificial intelligence and of mechanics.

Martian meteorites

To end this martian saga, which has taken us so far in terms of both time and space, did you know that we may already have found pieces of Mars on Earth? Have you heard of the SNC (Shergotty-Nakhla-Chassigny) meteorites? These are meteorites that come from Mars: at present nine of them are known. In the last decade, oxygen isotope analysis and the ratios of trace elements have shown that, of the several thousand meteorites available to us, a small group has actually come from another celestial body. In addition, the crystallization ages of these meteorites are astoundingly low: they lie between 160 million and 1.3 billion years, whereas all the other meteorites crystallized 4.5 billion years ago, at the time when they were actually formed. The recent date for the SNC meteorites implies that their parent body was not an asteroid, because none of those could have undergone any significant melting at such a recent date. They cannot come from the Moon, because its youngest rocks have ages of more than three billion years. The parent body is therefore the size of a planet, and is, moreover, a terrestrial one, because the giant planets are all gaseous.

Inclusions have been discovered inside SNC meteorites in which the argon and xenon isotope ratios are characteristic of those that have been measured in the martian atmosphere. It seems that the SNC meteorites have come from there! In addition, their chemical composition is similar to that of the soil near the Viking Landers. Finally, most of them have been found to exhibit shock effects, probably produced by impacts, some of which have been dated to less than 30 million years ago.

The questions persist, however: how did these small pieces of Mars reach the Earth? The best explanation is that, following a relatively recent impact on Mars, fragments of the surface were ejected into space, where they orbited as tiny individual planets around the Sun. Some of them eventually finished by falling onto the Earth.

In an attempt to go even farther, people have scanned the photographs taken by the Viking Orbiters for elongated impact craters. To be launched into space, these fragments must have been hurled from the surface by an impact at a low angle of incidence. The most extraordinary thing is that P. J. Mouginis-Mark from the University of Hawaii, has detected eight such craters, all in the Tharsis volcanic region. The most interesting site is at the foot of the northern slopes of Ceraunius Tholus, which is a volcano 120 km across at its base, and 6000 m high. This impact has left an oblong crater, 18 km by 34 km, which is, moreover, lobate, indicating the presence of permafrost in the subsoil. This interesting collection of features seems to be reinforced by investigations of the amount of volatile compounds in SNC meteorites. In 1990, J. L. Gooding of NASA's Johnson Space Center, concluded, from his measurements, that geochemical processes must have taken place in an aqueous phase on the parent body, oxidizing the minerals to form carbonates and sulfates. Yet again, we find powerful evidence for a link with the presence of water.

Will this incredible 'espionage' affair stop here? Using ^{13}C and ^{18}O isotope ratios from SNC meteorites, like those Schidlowski determined for stromatolites, M. V. Ivanov, from the Institute of Microbiology in Moscow, has been able to draw a correspondence with his own studies of methanogenesis, where metabolism by the methanobacterium *Formicum* converts carbon dioxide and hydrogen into methane. He concludes that such microbial methanogenesis could account for the isotopic ratios found in the SNC meteorites. This sensational news, which needs to be confirmed by other laboratories, would be of crucial importance in confirming the existence of life on Mars, to such an extent that, at the workshop on martian simulation held at Bad Honnef in 1992, L. M. Mukhin from the Space Research Institute in Moscow, suggested that Ivanov's paper should be entitled 'First evidence of life on Mars'!

PART II
Extraterrestrials

5

Intelligence

With our discussion of Mars and our hopes of finding traces of primitive life there, we leave the fourth stage of our search for life in the universe, that of the primitive biological stage. We are now about to embark on the last stage, a search for 'advanced' life forms, and intelligence. We have seen how true life forms appeared as the result of a chain of steps that have not yet been fully elucidated. The steps are, in effect, those that define life as it occurs on Earth. The same applies to intelligence. Let us start with the most advanced, and the only form that we know: human intelligence. After all, this is where scientists have begun in implementing SETI, even if we take into account less highly evolved forms, such as the intelligence of dolphins or even of parrots, or parallel types, such as the artificial intelligence that we are trying to develop for the martian rovers.

Nevertheless, we should remember that human palaeontology tends to reveal the various stages in the evolution of our technology rather than of our intelligence. For SETI, the situation is similar: we hope to detect signs of intelligence through the intervention of technological developments. SETI is, as yet, merely the search for extraterrestrial technology. Consideration of our ancestors is therefore able to provide important information that will help us to define the scope of our investigation.

The emergence of the human species[*]

A revolution in the construction of our ancestral tree came with analysis of the sequence of nucleotides in the DNA of modern

[*] I base this chapter on a talk given in 1986, at Darwin College, Cambridge, by David Pilbeam, Professor of Anthropology at Harvard University.

hominoids: chimpanzees, gorillas, humans, orang-utans, and gibbons. As a result, we now believe that the gibbons branched off
ten million years ago, followed by the orang-utans, and that it was
only five million years ago that the chimpanzees and humans diverged. What was particularly surprising was to find such a close
parentage, both in time, and genetically, because we differ from the
chimpanzees in only 1 % of our genetic material. Like them, we
are African apes! As for the gorillas, they are also close relatives,
because their line diverged slightly more than five million years ago.

It is by studying the behavior of the great apes that we have made
most progress in understanding the earliest human ancestors, *Australopithecus*. They were the first to walk upright, and lived in East
Africa. In the form *Australopithecus afarensis*, they survived until
three million years ago. The line then splits into an *Australopithecus
africanus* branch, which lasted until two million years ago, followed
by *Australopithecus robustus*, persisting until one million years ago;
and finally a *Homo* branch. First we find *Homo habilis*, who died
out 1.6 million years ago, then *Homo erectus*, who lived as recently
as 300 000 years ago, and finally, *Homo sapiens*, which brings us to
the present. Initially *Homo sapiens* occurred in an archaic form, of
which the best-known representatives are the Neanderthals, found
in Europe and western Asia. Between 300 000 and 130 000 years ago
they were gradually replaced by a more evolved type, of which we
are part. Neanderthals and *Homo sapiens sapiens* existed alongside
one another, just as the final form of *Australopithecus* did alongside
the earliest representatives of the *Homo* line, and just as we still
live alongside our more distant relatives, the great apes. Unless we
protect them, they will also die out, and we shall be left alone, like
simple-minded idiots, sitting among smoldering ruins, and weeping
for the cousins we have murdered.

The behavior of our ancestors

As far as SETI is concerned, it is essential to determine the ways
in which our various ancestors behaved. This is the best way of
understanding how technology and then intelligence were able to
appear. First, according to David Pilbeam, we need to define the
principal characteristics that differentiate humans from the great

apes: an upright stance; bipedalism; freeing of the hands; the fabrication of tools; a large brain; complex, and above all acquired, cultural behavior; language; the adoption of a hunter–gatherer way of life; the use of a fixed base, from which expeditions are made daily, and where varied activities take place; and finally, control over the environment, and particularly the use of fire.

Australopithecus was probably the hunted rather than the hunter, and their behavior, both individually and socially, was like that of the apes. They lived on the savannas and in wooded terrain, rather than in the forests. Bipedal, they remained capable of climbing trees for foraging, protection and to rest. They were also capable of wandering from one source of food to another, with a far greater ease of locomotion than chimpanzees. Their food was based on plants, above all fruit, and dried grains. These vegetarians lived in small groups consisting of females and infants and an adult male. They used stones and pieces of wood to obtain food, and for the purposes of aggression, as do chimpanzees. They were probably incapable of language. In this respect, further knowledge about communication between chimpanzees under natural conditions would be extremely valuable. Why did they become bipeds? It was probably a result of living in a savanna habitat that forced them to undertake longer journeys to find food and sexual partners, to defend their territory, and also to carry food to their dependants.

In *Homo habilis* we find signs of improvement: the legs are longer, and the back is more flexible; base camps appear, where crude tools, like broken pebbles, are manufactured; meat enters the diet, both in the form of carrion and as small game. *Homo habilis* left Africa and spread into western Asia.

Homo erectus had a brain twice the size of that of *Australopithecus*: they entered into competition with the carnivores for meat, either as scavengers or perhaps even as hunters; they had mastered fire; and their tools are more regular in shape, which implies more highly evolved manipulative and mental skills. The technology, however, remained practically the same for a million years, which is a sign of an astonishingly stable pattern of behavior.

Neanderthal man had a greater diversity of tools, based on finer working, such as lances and knives. Thanks to their massive bodies and powerful muscles, they were capable of a great deal of activity.

They had stamina for hunting and were able to capture small game species either by speed or by wearing them down. From being scavengers, they became hunters.

Finally, the transition to modern humans took place 35 000 years ago in Europe, and between 200 000 and 100 000 years ago elsewhere. This brought with it painting, sculpture, language, sophisticated tools, the organized hunting of large game, and permanent habitations where extremely diverse activities were carried out. These allowed an increase in the population, both in numbers and in extent. *Homo sapiens* has conquered the planet and transformed it. The crucial stage was reached 10 000 years ago, in the Neolithic, when agriculture and the domestication of animals were introduced.

As David Pilbeam said: 'There is no single point at which we became human. Many "human" qualities appear to have developed extremely late ... The process of becoming human involved numerous individual steps – language, bipedalism, meat-eating, the use of fire, short pregnancies – all of which must be rather unlikely, and certainly not inevitable, events.' In view of this, should we not be very restrained in the ideas that we may have about intelligence and the forms that it may take, and also to be prepared for extraordinary discoveries if we should ever detect signals from extraterrestrial intelligences? The human example shows that the development of intelligence is not a linear process. It is easy to say, after the event, that a miracle came about by means of a grand plan. Like the development of the universe after the Big Bang, and the evolution of life following the end of the ancient bombardment, we must recognize that the development of intelligence took place in unexpected ways that were advanced or eliminated by the selection of the fittest.

Plate tectonics and intelligence

A good example of the frequently unexpected twists and turns of fate is given by the African origin of the human race. According to the paleontologist Yves Coppens: 'It appears more and more clearly that *Australopithecus* is the ancestor of *Homo* and that *Australopithecus* is well and truly African ... Practically the whole evolution of the

hominids is illustrated between the Rift Valley and the Indian Ocean
... together with the data that are required to show our relationship
to the gorillas and the chimpanzees.' But why should it be Africa,
and why that region in particular?

Some ten million years ago, the equatorial forests extended right
across Africa, from the Atlantic to the Indian Ocean. Then up-
welling of magma from the interior of the planet produced the giant
fracture zone of the Rift and major uplift of the regions bordering
it. This disrupted the rainfall regime, leading to the disappearance
of the forest on the eastern side. A climatic change that modified
the nature of the landscape was therefore directly responsible for
the emergence of hominids. 'Gorillas and chimpanzees may represent the descendants of those of our ancestors who survived in
forested country; whereas australopithecenes and humans may be
descended from ancestors who, isolated by a tectonic accident that
gradually became an ecological barrier, were confronted with a climate that steadily worsened, and who therefore had to adapt to an
open countryside.' Is this not, yet again, a wonderful, extraordinary,
and totally unexpected twist of fate?

Homo erectus and SETI

Yves Coppens has remarked that 'The bifacial axe, which was invented by *Homo erectus*, took on an increasingly important role,
which it was to maintain for a million years ... Although this tool
underwent changes through the millennia ... we must not forget
that this advance took place over ten thousand centuries!' The slow
rate of this evolution – or, to look at it another way – the duration of the technology, has been particularly well documented by
the systematic excavation, from top to bottom, of the Choukoutien
cave site, near Beijing, undertaken since 1987 by Chinese scientists.
The site was occupied by a subspecies of *Homo erectus*, *sinanthro-
pus*, between 460 000 and 230 000 years ago; 20 000 stone tools have
been analyzed. They form a simple tool-kit, based on flakes that are
worked to a greater or lesser extent to produce choppers and scrap-
ers. The material gradually changes from sandstone, which is easy
to work, to predominantly quartz, and then partially to flint, which

is more difficult to handle. However, this is a relatively modest form of evolution.

In fact, the only real progress recorded concerns the size of the tools; they went from an average weight of 50 g and a length of 6 cm, to less than 20 g and 4 cm. But these people already had a fairly respectable way of life. They used fire from the outset, had an energy-rich diet of meat, and lived in well-appointed caves. Such a low rate of evolution over such a long period of time, amounting to 200 millennia – or if you prefer, 2000 centuries – gives one food for thought ...

To my mind, such a fact is of considerable significance for SETI, because of the conception we may have about the duration of the civilizations that we are seeking. This duration is a fundamental factor affecting our chances of success, because the longer a civilization survives, the greater the probability of detecting it – if it is detectable at all. Our current pace of technological development is measured in terms of a generation or, say, between 10 and 100 years. Such a short time-scale does not help us to imagine civilizations that last for a million years. But the Beijing cave is there to show us that it is possible. When arguing for such a long duration among extraterrestrial civilizations, people have frequently resorted to suggesting that they might lose interest in progress, or take a specific decision in favor of zero growth. It is also possible to imagine a regression from an acquired culture back toward instinctive ideas, thus bringing intelligent technological evolution to a halt, or a reversion to a much slower genetic evolution, such as has occurred, for example, with the construction of nests or termite mounds.

A culture may, however, also reach an insurmountable conceptual ceiling. After all, the idea that better tools could be manufactured was probably beyond the comprehension of *Homo erectus*: their brains simply lacked the corresponding 'pigeon-hole' or 'bump' that their successors, the Neanderthals, possessed a long time afterward. Nevertheless, they provide SETI with an example of a civilization that extends over a period of a million years; a primitive one, it is true, but still one that was intelligent and technologically aware.

Mammalian intelligence

We would like to have far more details of the evolution of human intelligence than we have been able to gain from the limited evidence at our disposal illustrating the evolution of technology. Luckily, serious controlled studies of animal intelligence have been made over the last 20 years. Methods have been developed from *in situ* observations of animal behavior that aim to avoid any possible anthropological bias that would falsify the study of other species. The main areas of difficulty involve the recognition of signals and the way in which they are decoded.

In this respect, dolphins provide some fascinating insights, because they exhibit such complex, flexible behavior, as well as being capable of symbolic communication. Yet their evolutionary past is very different from that of humans. Our path to intelligence is not the only one, and this is of fundamental importance in SETI investigations.

Dolphins have large brains, advanced social behavior, and exchange signals that are not only vocal, but also visual, tactile, postural, and may perhaps even involve taste. Using a keyboard with nine keys, each of which represents a familiar object (a ball, a ring, etc.) and emits a different sound, Diana Reiss, of State University, San Francisco, has taught them to obtain these objects by making the same sounds themselves, without using the keyboard. Dolphins are thus able to learn and use new codes of communication. Such intelligent behavior, which involves both genes and the environment, has an adaptive value and enables them to react to new situations, to make use of past experiences, to form concepts and generalizations, and thus to respond rapidly and fully, without having recourse to a series of fruitless trial-and-error experiments.

The intelligence of birds

Similar faculties, which were thought to be the sole preserve of humans and perhaps a few nonhuman primates, have been demonstrated even in birds. Irene Pepperberg, of the Department of Anthropology at North Western University in Illinois, has been studying the behavior of Alex, a grey parrot from Gabon in Africa, since 1977. Alex, who was born the year before, lives free in the laboratory but is put in his cage at night, and has water and seed available at all times. As a reward for learning, he is given fruit, vegetables, and 'toys.'

Irene Pepperberg has developed three methods: Alex witnesses an illustrative training session between two of the researchers; a word is repeated to him in varying contexts; or, finally, if he makes a mistake, the situation is turned to good account by teaching him a new symbol. The results are astonishing, and are nothing like the tricks obtained with an 'intelligent' animal in a circus. Alex has learned the name of 35 objects or actions, of seven colors, and five shapes; he uses phrases like 'come here, I want x, and I want to go to y. He knows how to use the word 'no.' He makes combinations of words to identify, to request, to refuse, to classify, and to count. He understands concepts such as categories, resemblances, and differences – all of which reveal a capacity for abstraction. He also knows how to search for something following instructions.

Alex therefore possesses complex cognitive facilities, despite the fact that his cerebral organization differs considerably from those found in terrestrial and aquatic mammals. It therefore seems possible that there may be other forms of intelligence, developed for specialized tasks that we have yet to discover. Another point that is very relevant for SETI!

Technology without intelligence

On how low a level might we expect to find indications of intelligence? Among the social insects, bees use dance to communicate information about the position of a source of food. In addition, they are able to learn, as has been shown by the experiment where the source of food was moved every day by the same amount: they eventually went directly to the extrapolated position, rather than to the one that applied the day before. D. M. Raup, a professor at the University of Chicago has taken an extreme view, pointing out that the climate control of termites, and their constructions, the electrical signals in certain fish, and the magnetic navigation of certain birds, are examples of animal mechanisms that give functional results similar to those of humanoid intelligence. But they are all genetically based, the result of a long Darwinian evolution. Behavior that appears intelligent is not restricted to intelligent organisms, so that one could, according to Raup, speak of nonconscious intelligence. The term 'nonintelligent technology' may be preferable. We can, in any case, envisage the possibility of nonintelligent species using radio techniques for communication. This has not been shown to exist on Earth, but it is by no means impossible. Such a view can greatly expand the field of SETI: such intraspecies communications could last a long time, given that they arise through Darwinian selection, and could also emit significant levels of power, if such species were particularly prolific. We should remember the principle of biological saturation invoked by Schidlowski in connection with stromatolites.

The precursors of awareness

As far as the immediate precursors of true awareness are concerned, there are (for example), the famous experiments carried out in 1915 by W. Köhler on chimpanzees. Confronted with the problem of a banana hanging from the ceiling, with a box and a stick at his disposal, the chimpanzee is initially frustrated, and becomes agitated. Suddenly, his face changes: he immediately places the box underneath the banana, jumps on top, and seizing the stick, obtains the fruits of his brain-wave. This exhibits introspective behavior that is very close to the threshold of conscious, human thought patterns. There is no need to make haphazard trial-and-error experiments: all that is required is to simulate the problem in one's brain. According to W. H. Calvin, from the University of Washington in Seattle, such a technique allows complex scenarios to be explored. What is needed is a memory that acts like a buffer, capable of storing random sequences of scenarios, and also a memory that recalls transitory past experiences. By comparing the contents of the two, it is possible to classify the scenarios according to their merits for solving the problem.

The actual neural machinery involved is enormous. It must have progressed through Darwinian evolution and may have arrived at a state of perfection by the time hunting with thrown projectiles first appeared, thus giving our ancestors a distinct advantage. Throwing a stone requires the activation of more than 80 muscles in a planned sequence, with a time resolution far finer than the normal reaction time, which is reckoned in tenths of a second. The whole process of the throw needs to be mapped in detail, rapidly and completely, just before the critical actions are triggered, otherwise the stone will not hit the target.

The process was carried out by what Calvin calls the 'Darwinian machine', which was also important in controlling the hammering action that is required for the fabrication of tools. When it was resting, the 'machine' was available for arranging concepts, or symbols, or words, in sequences that opened up possibilities for future advances in our intelligence. Calvin notes that the potential for throwing and hammering actions did not appear with the beginning of upright stance and the liberation of the hands, but two or three million years later. It blossomed during the Ice Ages, a series of climatic oscillations each lasting about 100 000 years, which subjected *Homo habilis* and *Homo erectus* to intense selective pressure that favored major progress among the survivors. Even more rapid events, on a regional scale, may have occurred as a result of sudden changes in ocean currents. In Europe, for example, 11 000 years ago, the temperature rose by 7 deg.C and rainfall by 50 % in just 10–20 years. The selection of versatile behavior that ensured the provision of food, of shelter, and the survival of offspring – all tasks that were possible thanks to the 'Darwinian machine' – were thereby favored.

Yet again, we see the influence of various chance events that have affected human intelligence, and glimpse the hitherto unsuspected other paths that might have occurred elsewhere in space. This represents a truly vast horizon for SETI to scan.

The footprints of *Australopithecus*

Nevertheless, the most significant witness of human development, at least in the informed opinion of numerous palaeoanthropologists, is shown by traces of footprints left behind by the first hominids to have walked upright. In 1978, Mary Leakey, who has devoted her entire career since 1935 to excavations in East Africa, discovered at a site named Laetoli in Tanzania, three sets of tracks, 25 m long, left by australopithecenes in a layer of ashes deposited by the volcano known as Sadiman, 3.7 million years ago.

The exceptional state of conservation is because of an improbable sequence of chance events, which is described by Don Johanson, who, with Yves Coppens and Maurice Taieb, was the co-discoverer of the *Australopithecus afarensis* skeleton known as 'Lucy': 'Sadiman had to blow out a particular kind of ash. Rain had to fall on it almost immediately. Hominids had to follow on the heels of the rain. The sun had to come out promptly and harden their footprints. Then another blast from Sadiman had to cover and preserve them before another obliterating shower came along. ... Indeed there had to be just what the beginning of a rainy season produces: sporadic showers interspersed with intervals of hot sun.' The left-hand track was left by a small australopithecene, and the one on the right, which was probably double, was produced by a large individual, followed by a medium-sized one. We imagine that on that remote day, a female australopithecene was following a male, while a child attempted to keep up and take large steps like its parents. A touching, mute witness to family life long ago.

Between these first steps taken by our ancestors, and our own first steps on another heavenly body, the Moon, 3.7 million years have elapsed; in other words, one-thousandth of the age of life on Earth. In scientific terms, one-thousandth is a very small difference. Our immediate reaction is that we could reasonably try to extrapolate conditions another one-thousandth part into the future.

Given the extraordinary advance that intelligence has made between *Australopithecus* and Apollo, one conclusion is inevitable: it would be completely unreasonable to imagine that human intelligence is the pinnacle of achievement throughout the universe. Anyway, where will we be in three million years? Unfortunately, our knowledge is still incapable of offering us any guidance: we can only speculate. Yet we still want to know. It is here that SETI might enlighten us as to the possible futures that may await us. It is, in fact, the only observational method that is capable, by harvesting information about possible extraterrestrial civilizations, of throwing any light on the all-important question of our future existence.

Subconscious opposition to SETI

But we must recognize that any such evidence would still be very poorly received, even by many scientists and, it would seem, particularly in France. I have frequently encountered a gut reaction opposed to the very suggestion of the existence of possible intelligences superior to our own. (After all, for centuries the human male, full of his own superiority, denied intelligence in animals, and even in women.) If, impressed by the forces of nature, such individuals occasionally acknowledge the existence of infinite intelligences, the latter form part of a world outside the real one, and are thus beyond competition. If anyone dares to speak of intelligences that are possibly superior to their own, living in the real universe, all such people do is dismiss the idea with a shrug of their shoulders!

Such behavior may lie at the basis of the opposition held by some people to the very idea of extraterrestrial intelligence. I hope that we can take a dispassionate, subjective view of the primitive intellectual response that may become involved in this field of research. Motivated as we are by irrational behavior, we should not try to ignore this feeling.

Kardashev's supercivilizations

Nearly 30 years ago, in 1964, a young Soviet student who was preparing his thesis on radio astronomy, was not frightened to imagine intelligences superior to his own. Since then, Nikolai Kardashev has become director of the largest radio telescope in the world working at millimeter wavelengths, a paraboloid 70 m in diameter, which is under construction near Samarkand. Starting from an appraisal that is still valid, he noted that human consumption of energy was at a level of about 10^{13} watts (10 000 gigawatts), and had been increasing at a rate of several per cent per year for 60 years. Limiting ourselves to an annual increase of just 1 %, it follows that, at this rate, in 3200 years it will reach the total energy radiated by the Sun, and, in 5800 years, the amount emitted by the whole of our Galaxy. At its current rate of increase of 10 %, the amount of information handled by our society would increase, in just 2000 years, by a factor of 10^{80}. Expressed in binary digits (bits) it will far exceed the total number of atoms in the observable universe. Such a vast quantity of information could not, therefore, be stored in any material form of memory.

The same considerations apply to population: based on the current usage of ten tonnes of material per person, at an annual rate of increase of 4%, the population will require, in 2000 years, the total mass of ten million galaxies. It is staggering to realize that human activity, extrapolated over a human time-scale, could involve activity on a cosmic scale!

Kardashev draws two forms of conclusion from this. First, in searching for extraterrestrial civilizations, we should not close our eyes to the possibility of the existence of supercivilizations, which he classifies into three types: I, those that have energy requirements comparable with the amount radiated by their sun and intercepted by their planet (10^{13} watts); II, those that use the energy output of their sun (10^{26} watts); and III, those that use the energy emitted by a galaxy (10^{37} watts). Second, Kardashev concludes that the exponential progression that is currently advancing our civilization will, inevitably, be restricted, and that our current expansion is a transitory phase. In short: it can't last long!

What will happen? The conquest of space can, of course, provide us with living space, energy, and materials. But it cannot be more than a short episode. The expansion of our civilization in space cannot take place faster than the speed of light, and, because of this, in 1000 years' time, its growth will cease to be exponential, and will become slower. (It will become proportional to the square of the elapsed time.)

According to Kardashev, supercivilizations that are far less insatiable and prolific, but far more advanced than our own, can, nevertheless, still exist. This is why he has recommended searching for the various losses that such 'astro-technologies' may involve, in the form of emissions – such as infrared radiation – from cosmic-scale engineering projects, or even of observing such works directly by means of their end results. What an extraordinary prospect! ...

In addition, our current stage of civilization is extremely transitory on a cosmic scale. What are a few millennia in comparison with billions of years? As a result, there should be other 'civilizations' that are less advanced, such as the bacteria that ruled the Earth for billions of years, or the dinosaurs for 100 million years – and thus difficult to detect. In contrast, there should also be others more advanced, and thus potentially detectable. But civilizations similar to our own are unlikely.

This physicist's point of view needs to be complemented by those involving other disciplines. One factor that has been mentioned is the conceptual ceiling that may have hampered *Homo erectus*. Without the emergence of the Neanderthals, they might have been able to continue to avoid exponential growth for a very long time. Moreover, such a ceiling might be a general trait: are we not, in a certain sense, rather presumptuous, if, in searching for advanced intelligences, we do not set an upper limit to the level of intelligence possible in our – and I stress 'our' – universe? We should not forget that in Linde's chaotic Big Bang cosmology, our universe may merely be a specific case that is not particularly remarkable, and that if we want to find intelligences that are infinitely superior to our own, we need to consider an indefinite number of other universes. For their part, sociologists may invoke a nuclear apocalypse or pollution as examples of suicidal behavior that impose their own limitations. Degeneration or the loss of interest by a civilization are also other possible limiting factors, or, at our particular stage, the destruction of the Earth through collision with an asteroid or comet, or the planet's ejection from the Solar System as a result of chaotic perturbations by the other planets. We should not forget that the very fact that we have arrived where we are today is actually a miracle, and that our lives are, in truth, a cosmic odyssey.

6

The SETI pioneers

One morning, I said farewell to my father on one of the platforms of the Gare Saint-Lazare in Paris: we waved to one another as long as we could in a protracted goodbye. I was embarking on a great adventure, and did not expect to return for two years. My immediate destination was Le Havre, and there the steamship *de Grasse*, and a week's voyage. Out on the immense ocean, under a starry sky, I could see phosphorescent eddies: the light of microscopic animals glowing beneath that of the stars, one responding to the other across gulfs of billions of years in time, and over thousands of light-years in space. I was threading my way between sky and sea, toward America, the vanguard of the science of the future.

My destination was the prestigious Cornell University, where my director was sending me for two years. Louis Leprince-Ringuet, Professor of Physics at the École Polytechnique in Paris, and founder of the cosmic-ray laboratory, was a great patron, whose nature and behavior were both unconventional. He made a point of sending his young colleagues out as early as possible to rub shoulders with other people elsewhere. At Cornell, beside Lake Cayuga, there was no SETI. Carl Sagan had not yet established his laboratory there. Working there, however, was Hans Bethe, who was awarded the Nobel prize for Physics for having discovered the source of the Sun's energy: the famous carbon cycle. 'Le Prince' sent me to him, because in our investigations of cosmic rays, we had discovered an intriguing problem in nuclear physics. Once there, I also helped the Cocconis, whose observations were being conducted beneath the deep waters of the lake, where they were attempting to detect mu mesons. Every Wednesday, a gathering of the Journal Club was held, where the dazzling, passionate, avant-garde Phil Morrison took our breath away with his staggering visions of the future.

Initial hypotheses

Guiseppi Cocconi and Phil Morrison are the two men who, in a historic article in *Nature*, first showed that it would be possible, using the new techniques of radio astronomy, to communicate across interstellar space with possible extraterrestrial civilizations. In fact, I missed the event, because their famous article appeared eight years after I had returned to Mont Sainte-Geneviève in Paris, and at the very time when I was moving to Meudon. I had been attracted by the completely new field of radio astronomy, which, in France, was taking shape under the active, enlightened, dynamic, and occasionally even humorous, direction of Jean-François Denisse.

Cocconi and Morrison calculated that if other radio astronomers elsewhere in the universe had radio telescopes and receivers comparable to those we had in 1959, and radiative powers similar to those that we had at our disposal, it would be possible, despite the colossal distances between stars, to exchange radio signals, and thus communicate. From the whole vast spectrum of possible wavelengths, they also recommended using the 21-cm emission from hydrogen atoms. According to them, because hydrogen is by far the most abundant element in the universe, the 21-cm wavelength, which is physically conspicuous, could serve as a universal standard for the community of galactic civilizations.

The first listening session

That same year, a young American, Frank Drake, who was working on his doctoral thesis, and following an experimental rather than theoretical line of inquiry, suggested that he should build a special radio receiver to listen for possible signals. His supervisor, Otto Struve, who came from a family of eminent astronomers, backed this proposal, and placed the newly built 24-m radio telescope at the National Radio Astronomy Observatory at Green Bank in West Virginia at his disposal. Drake had also evaluated the potential of radio-astronomical techniques for interstellar communication, quite independently of Coccioni and Morrison. When the time is ripe, an idea often arises simultaneously in different places.

Drake took a year to adapt his receiver to the region around the 21-cm wavelength, and selected as his targets, two of the closest stars that most resemble the Sun, Tau Ceti and Epsilon Eridani. The relative proximity of these stars, and the possibility that they possess terrestrial-type planets favorable for life, increased his chances. At the time, no one had any precise indications of the existence of such planets, but in this connection it is amusing to hark back to words written by the great French astronomer, Camille Flammarion. In his *Les Étoiles et les curiosités du ciel* ['Stars and curiosities of the sky'], published in 1882, he said: 'The star tau in the Whale is noticeable because of its rapidity ... It moves as if it were traveling across the immensity [of space] in conjunction with us, but faster than us. The sidereal populations that inhabit this star's system may perhaps be bound to us in an eternal destiny. It would be of the very greatest interest to measure the [distance] of this star.' Prophetic words that capture the imagination and are an invitation to adventure. A century later we know that Tau Ceti lies 12 light-years away, and that the similar star Epsilon Eridani is a good candidate for possessing a planet.

For a few weeks, Drake carried out his listening program, known as OZMA, which was the first such program to be directed at the stars. First star, nothing. Despondency. His receiver was very primitive: it had just a single channel, like all others at the time, which did not permit effective exploration of wavelengths around 21 cm to detect any possible emission. ... Second star: Bang! The pen-recorder hit the stop because of the strength of the signal!

With adrenaline racing, Drake had an immediate, profound, and notable reaction: Could it really be true that all it took to make such an enormous discovery was for a student to spend a year modifying a radio-astronomy receiver? Was it really so easy? Why did no one think of it before? What incredible possibilities were there that might have gone unnoticed for the lack of such a tiny effort? Drake kept his head, however: it was too simple to be true! But more of that later.

It was a great disappointment, of course, but the basic method was to hold good for decades. For a first attempt, it was something to have detected an artificial radio signal coming from stratospheric 'intelligence' (using the word in the extended, American, sense). Frank Drake had become history. Whoever, in future, does discover a real, artificial, extraterrestrial signal, even if it is centuries hence, will probably pay tribute to his work.

The three basic hypotheses

Thirty years after the first theoretical and experimental attempts by Cocconi, Morrison, and Drake, SETI has become a hundred-million dollar program in the United States. But there are still numerous critics: witness the fact that, in 1994, the US Congress withdrew all funds for NASA's SETI project. So it is important to lay out the arguments for it in full.

Like every serious scientific study, SETI proposes certain reasonable working hypotheses, draws conclusions from them, then attempts to verify them through observation or experiment, before proceeding further. This method of procedure avoids the dangerous reefs of gratuitous speculation, which are all too often both pointless and harmful.

First hypothesis: 'Life on Earth is the result of natural evolution by physical processes that apply throughout the universe.'

The developments reported in the first part of this book illustrate that this hypothesis is entirely reasonable. 'Physics' is taken in the broader sense that includes all the natural sciences – chemistry, biology, etc. In any case, a hypothesis does not mean a dogma, or an axiom, or a principle, or even a revealed truth. When I was working in cosmology, I detested the famous principles that some people evoke, such as the 'perfect cosmological principle.' Why should nature follow, *a priori*, certain specific *diktats*? Although a hypothesis may appear to be valid whenever we are able to test it, that does not imply that it is actually a principle. Frequently, these false assumptions, which are no more than well-established facts (which may, just possibly, be elevated to the status of a law), are overturned by a new experimental or observational investigation. That nothing can exceed the speed of light is well established experimentally, and has been incorporated in an excellent theory,

Einstein's Theory of Relativity. Yet, under certain circumstances, fluctuations of quantum physics allow this limit to be exceeded, albeit minimally. Although our understanding of the universe is powerfully upheld by mathematics, nature is not simply a set of mathematical equations.

Second hypothesis: 'What has happened on Earth could have happened elsewhere.'

There are, on average, billions of stars in every galaxy. In the observable universe, out to the cosmological horizon at a distance of some 15 billion light-years, there are 100 billion galaxies. Among the 100 billion times 10 billion (or 10^{21}) stars, 10% are similar to the Sun. In addition, the Earth has existed for only 4.5 billion years, whereas the universe began with the Big Bang some 15 billion years ago, and most of its galaxies have existed for at least 12 billion years, roughly three times the age of the Earth. There is therefore no lack of potential 'elsewheres.'

Third hypothesis: 'Human intelligence is not the ultimate, the *ne plus ultra* of what the universe could produce.'

It has taken 4.5 billion years for us to arrive at the stage that we currently occupy. But other stars similar to the Sun have existed for much longer. (At very distant epochs, however, because the stars did not contain any elements heavier than helium, they could not have had terrestrial-type planets.) A reasonable estimate suggests that the oldest stars (with heavy elements) and similar to the Sun existed ten billion years ago. Such stars therefore have a head–start of some five billion years. Remember that the vast evolutionary progress between *Australopithecus* and ourselves took just 3.7 million years: what is that compared with five billion? One part in 1300! How can we reject the idea that in billions of years, in billions of galaxies, each containing billions of stars, physical evolutionary processes may have produced more advanced results than those found on our small globe toward the end of what we insignificantly call our 20th century? No one can, *a priori*, reasonably reject this possibility.

Possible consequence: 'The universe may contain more advanced forms of life than ourselves.'

Whether they are little green men, or Kardashev's supercivilizations; whether they are of planetary and biological origin, like us, or arose in interstellar space through the interaction of nebulosity and magnetism, like Fred Hoyle's *Black Cloud*, or are the product of artificial development based on a more highly evolved computer technology than our own; whether they correspond to societies consisting of multiple organisms or to single organisms that we would call 'beings', 'races', 'societies', 'technologies', 'civilizations', 'colonies', 'persons', or 'individuals', we know absolutely nothing about them. The only characteristic that we can attribute to these advanced forms is that they are extraterrestrial. Let us, therefore, simply call these 'others': 'extraterrestrials.'

Observational test: 'Undertake SETI'

SETI is, in fact, the only method at our disposal of trying to verify, by observation, that extraterrestrials exist. We can try to pick up their emissions, whether intentional or not, by means of electromagnetic waves, which are information carriers that are relatively accessible to our current technology, and which propagate at the maximum possible speed, that of light, 300 000 km/second. If we were to receive any such signals, and were able to prove that they were artificial, we would have shown that the various hypotheses and the possible consequence are valid, and, in addition, that we are not alone in the universe ...

Millions of extraterrestrials

How many extraterrestrials are there? Without hesitation, I would say that millions may exist. This number may seem excessive, but it is nothing. By way of example, let us imagine that there is just one extraterrestrial society for every 10 000 galaxies. This is an extremely small figure according to current opinion, which aims to search for another civilization in our own galaxy. Taken over the observable universe as a whole, however, such a rate corresponds, in effect, to ten million societies. If we reckon the number of individuals as 10 billion per society – as it will shortly be in our own, human society – we obtain, for the observable universe, 100 million billion extraterrestrials! In comparison, suggesting that there may be some millions of them appears extremely modest.

Ten thousand galaxies ...

Talking of millions of extraterrestrials does not mean that all we have to do is switch on a simple transistor radio to capture their messages, because, on average, we would have to explore 10 000 galaxies to have a reasonable chance of finding one that is 'inhabited.' Study of the galaxies in our neighborhood shows that, in general, they occur in groups or clusters. The Galaxy, for example, and the Andromeda Galaxy are two of the principal members of the Local Group, which consists of about a dozen galaxies within a radius of three million light-years. The great Franco-American galaxy specialist, Gérard de Vaucouleurs, Professor of Astronomy at the University of Texas, has cataloged about 15 similar groups out to a distance of 30 million light-years. We have to go much farther to reach the nearest clusters, which contain far more galaxies: the Virgo Cluster, with more 2300 galaxies, lies at a distance of 50 million light-years; the Coma Cluster, with a thousand galaxies, is at 300 million light-years, as is the Hercules Supercluster, which has some 10 000 members.

Studying 10 000 galaxies therefore requires probing space out to a few hundred million light-years, which is an enormous distance when compared with the tiny 100 light-years, which is the maximum distance of any of the 1000 target stars that were chosen for inclusion on the program that NASA initiated in 1992.

The Drake Equation

The problem of detecting extraterrestrials has been aptly specified by the famous equation suggested by Frank Drake. Mathematically, his very simple equation requires only a series of multiplications. It gives the number of civilizations based on a life form similar to our own, in just our Galaxy, that would be capable of communicating over interstellar distances:

$$N = R \times S \times P \times E \times L \times I \times C \times V$$

This contains cosmic, biological, and technological terms. The first expresses the conditions necessary for the existence of stars with planets suitable for the appearance of life:

R is the number of stars formed per year in our Galaxy;
S is the fraction of these stars that are of solar type;
P is the fraction of these that have planets;
E is the number of planets situated at distance from the star that are suitable to the appearance of life.

Next, we have the biological terms:

L is the fraction of planets on which life appears;
I is the fraction of these on which intelligence develops.

Finally, the technological terms:

C is the fraction of intelligent species that develop communications technology;
V is the lifetime of the communication phase in years.

These terms express, very simply, in numerical terms, the various, successive conditions necessary to fulfil the desired aim: i.e., the existence of civilizations similar to our own that could undertake communications.

For decades, ever since it was first proposed, people have struggled in vain to set specific values for each of these terms. Only the stellar formation rate and the fraction of stars that are similar to the Sun may be quantified. But we still have insufficient information about the presence of planets, and even less about the number that are appropriately located.

Because life arose quickly on Earth, L may be close to 1, but what value should we adopt for I, when we recall that our intelligence took something like four billion years to develop, which is a figure on a truly cosmic scale? C may also be close to 1, but what might V be? For us, the communication phase has only existed for about

30 years, and then only thanks to the impetus given by the SETI pioneers, and to developments in radio astronomy.

How long will we have the desire to communicate? If we do not disappear in some catastrophe in the next 100 years, will our technology persist for a million years, like that of *Homo erectus*, even though that, in itself, was insufficient? There are so many uncertainties that N may be taken to lie anywhere between ten billion, the number of solar-type stars in our Galaxy, and one, representing the only known civilization, our own. It is this impossibility of obtaining an answer that has launched scientists on new forms of observation: sophisticated techniques for attempting to discover planets around other stars, and, above all, undertaking SETI itself to try to detect actual extraterrestrial signals.

7

Why use radio waves?

Surely the best way of searching for extraterrestrials would be to go from star to star to see if one of their planets showed signs of intelligent life? Without necessarily having to go there ourselves, all we need do is to send one of the robots that in recent years have enabled us to explore the planets and satellites in our Solar System, all the way from Mercury to Neptune.

These televised wanderings, reinforced by all sorts of measurements and discoveries, have provided us with information about an impressive series of bodies: from Mercury to the satellites Phobos and Deimos; from the asteroid Gaspra and the nucleus of Comet Halley to Titan (which we have discussed in detail), and out to Triton, which also has an atmosphere of nitrogen. The latter, at a distance of 4.5 billion km, 30 times the Earth's distance from the Sun, was the final destination of these historic journeys. So why not carry on, and go at least to the nearest stars, Proxima or Alpha Centauri? Are we not already in the process of doing just that, given that the Voyager probes, and their precursors, the Pioneer probes, have left the Solar System?

The distance record is held by Pioneer 10, launched in 1973, which is now some six billion kilometers away. For some years, studies have been made for the Thousand Astronomical Units (TAU) project, which envisages a device that would reach 1000 times the Earth's distance from the Sun, out in interstellar space, i.e., 150 billion km, or 130 light-hours. Launched by conventional chemical rockets, the probe's journey would last 50 years, which is a span of time that humans are beginning to consider feasible for a scientific exploration. In 1993, Claudio Maccone, from the Colombo Center for Astrodynamics in Turin, and myself proposed a mission to ESA, in which a probe would be sent to a gravitational focus of the Sun, at 550 AU, using a solar sail. In this 'FOCAL' project, the tremendous amplification provided by the Sun's lensing effect would allow breathtaking SETI investigations. But – and this is the crucial point – Proxima Centauri lies at a distance of four light-years, i.e., 35 000 light-hours or 300 times the distance that TAU would cover in half a century! The trip would take 150 centuries! What scientist would have sufficient motivation to launch such an undertaking, and what politician would provide the necessary support?

Solar sails

Nevertheless, scientists and engineers have not been put off from attempting to find new methods. One of them was sketched out by Hermann Oberth, the pioneer of astronautics. In 1894 at the age of 30, he published *Die Rakete zu den Planetraümen* [Rockets to Interplanetary Space], where he laid the foundations of methods of propulsion by reaction in the vacuum of space. He was also one of the founders of the Verein für Raumschiffahrt [Association for Space Flight], from which Werner von Braun would later emerge. The VfR built the first rockets powered by kerosine and liquid oxygen, to Oberth's designs. In addition to this breakthrough, the latter also put forward the idea of propulsion by solar sail.

If a large, reflecting surface is exposed to solar radiation in space, every photon, as it is reflected, exerts a small pressure on the surface. For every hectare $(10\,000\,\mathrm{m}^2)$ of surface, the force exerted by this radiation pressure is 10 g. This is very small, but if the surface consists of a very thin film, such as aluminized Mylar, 10 microns thick, for example, it is possible to obtain a solid, low-mass sail that will capture the stream of photons from the Sun, and impart an acceleration that is by no means negligible. Such a light sail, placed in orbit high above the Earth, and thus free from any atmospheric braking, would acquire an ever-increasing speed: in a year it would reach the Moon. In addition, it would be so little affected by the Earth's gravity, that it would require only another year to reach Mars.

This fabulous method has given rise to a proposal for a 'regatta' in interplanetary space: three designs, one European, one American, and one Japanese, each weighing 200 kg, would be launched by an Ariane or Proton rocket, and race to the Moon. The first to send back a photograph of the far side would have won the first space regatta. The technological interest and challenge are certainly there, but, as yet, the required funding has not been obtained. Such projects pose interesting problems concerning the deployment in space of large, thin surfaces, a quarter of a hectare in area (say). This might be achieved by diagonal, telescopic or inflatable masts, or else by using a double skin, a sandwich construction, that would be inflated and would then become rigid through the action of solar ultraviolet radiation. The interest in such structures for future large radio telescopes in space, intended for SETI, is obvious.

The problems of maneuvring these sails are also fascinating; peripheral shutters that may be inclined at greater or lesser angles may be used to orient the sail in the 'wind.' Christian Marchal, Research Director at the Office national d'études et de recherches aérospatiales [National Office for Aerospace Study and Research], is investigating the use of magnetic effects. By placing a conductor around the circumference and then passing a current through it – the power being provided by solar panels – we could obtain a magnetic loop that would alter its orientation depending on the terrestrial or interplanetary magnetic field. With just a simple reversing switch, it would be possible to produce movement in the opposite direction, and thus change the orientation of the sail at will.

Even more grandiose plans have been proposed for voyages to the stars. A sail several square kilometers in extent could reach Proxima Centauri in just one or two decades, provided it was 'blown along' by a powerful laser beam. (In interstellar space, far from any star, it is just as dark as it is on Earth at night.) This laser beam would be produced from solar radiation by a power station orbiting the Sun close to the orbit of Mercury. The interesting point is that the probe would not carry any propellant, and would therefore not be weighed down by any mass that would later be ejected.

This technology is not likely to be introduced tomorrow. Nevertheless, the first step has been taken, using the former Soviet Union's Mir space station. In 1992, the French cosmonaut who flew on board deployed an experimental sail, of side 40 cm, to make an accurate test of its behaviour. At the end of 1992, in project Znamya, they took a further step with a sail 25 m square. Neil Armstrong's 'small step' on the Moon in 1969 proved that small advances are frequently 'giant leaps for mankind.' If the successors to the Soviet Union persist, perhaps we shall see sails several hectares in area deployed around the year 2000.

Using radiation

If, in this century, there is no question of our going ourselves or sending robots, which are difficult to accelerate, greetings are likely to arrive by means of radiation. This is how astronomy has operated for thousands of years: people look at the sky, photograph it, or record its light (or more correctly its radiation), at various wavelengths, whether as visible light or as infrared, radio, ultraviolet, X-ray, or gamma-ray radiation.

These electromagnetic waves are perturbations of electric and magnetic fields that propagate through space – the more readily, the closer space is to a true vacuum – at the velocity c (299 792 458 m/second), which, according to Einstein's Theory of Relativity, cannot be exceeded by any material body. In fact, energy must be imparted to any such body to accelerate it, and the closer the velocity becomes to c, the greater the amount of energy required. To attain c the body would have to be supplied with an infinite amount of energy. This is why photons (the particles associated with electromagnetic waves in the wave/particle view of quantum physics, and which move at velocity c) are considered to have zero mass when they are at rest. The velocity c is a physical constant in both relativity and quantum theories. It is also a universal constant, at least in the universe that we inhabit. One additional point should be noted: although light propagates at a velocity different from c in air, glass, etc., this is because of interactions with the atoms in the substance concerned.

Electromagnetic waves therefore allow us, in principle, to 'see' extraterrestrials from a distance, and in the quickest possible manner, because they are the fastest messengers.

Neutrinos

Electromagnetic waves are not unique, however, as a medium for carrying information. A new astronomy (i.e., a new method of seeing things at a distance), was born in a spectacular fashion on 23 February 1987. Twenty neutrinos were recorded in several ground-based laboratories at the very instant that a star collapsed and gave rise to the explosion of supernova 1987A, in the Large Magellanic Cloud, one of the closest, small, neighboring galaxies (but still lying at a distance of 165 000 light-years). Some of these laboratories had been recording neutrinos produced by the nuclear reactions taking place at the center of the Sun; 1987A's neutrinos arrived from a far greater distance.

Neutrinos have the smallest mass of any known particles – to such an extent that measuring their mass is still beyond the capacity of even our most sensitive instruments. They are thus able to propagate at a velocity close to that of light if they are given a reasonable energy, and are able to carry information rapidly across the depths of the universe. They, rather than photons, enable us to come closest to the cosmological horizon, because electromagnetic photons become trapped by the dense 'soup' of particles that occurred during the first instants after the Big Bang. With photons, we are not able to 'see' any farther back than 300 000 years after time zero, at a time when electrons and primordial nuclei combined into atoms, which were relatively transparent to radiation. In contrast, the initial soup presented hardly any obstacle to neutrinos; their observation is limited only by the instant at which they appeared in sufficient quantity, which was toward the end of the 'Longest Second' (p. 10). Unfortunately, precisely because their propagation is practically unaffected by any obstacles, it is very difficult to detect them through their interactions with the material in our instruments. Despite their great size, our neutrino detectors observed only 20 when supernova 1987A collapsed. The technology for exploring the universe with neutrinos is still in its infancy, unless, that is, extraterrestrial astro-engineering projects happen to produce them in quantity ...

Gravitational waves

In passing, I will do no more than mention cosmic rays of astrophysical origin (from the Sun, interstellar space, quasars, etc.), but I do want to say something about gravitational waves, which were predicted by Einstein. We are beginning to get an indication of their existence from our studies of double pulsars. These waves also propagate at the speed of light, c.

A new generation of very expensive instruments are being planned to try to detect gravitational waves systematically. When supernova 1987A collapsed, for example, it suddenly changed the curvature of space in its vicinity, and this alteration, in the form of a spherical wave, must have gradually propagated outward, causing a transient curvature of space as it did so, just as a stone thrown into a pond generates a wave of curvature on the surface. If one of the future instruments had been in service at the time, we might have been able to detect the wave at the same time as the 20 neutrinos.

Another method of detection, currently within our capabilities, involves pulsars. It is possible to detect all the radio pulses from a pulsar, without missing any, and we can time their instants of arrival to better than a microsecond. Imagine what happens when a gravitational wave crosses the path of the radio waves as they travel from the pulsar to us. When the waves encounter a region of space where the curvature varies, their path is forced to conform to the new, more complex curvature, and their arrival times will be slightly retarded or advanced. There are high expectations that this new technique will be able to detect gravitational waves. Because we cannot rule out the possibility that a supercivilization might be able to cause matter to collapse, by high-powered astroengineering, this might be one way of detecting such a civilization's existence.

However, let us be realistic. Currently, the most feasible method consists of trying to detect the electromagnetic waves produced by extraterrestrial civilizations. Because no known physical principle forbids the existence of these waves, simple pragmatism suggests that we should search for them.

Electromagnetic waves

Electromagnetic waves are all similar in nature, and they are defined by their wavelength. The perturbations in the electromagnetic fields may be represented, in a simplified case, as alternating variations in the strength of these fields at a given point. The strength increases, decreases, increases, etc., periodically, in a regular fashion; for example at 100 million times per second, when the period is obviously one hundred-millionth of a second. However, these regular variations at a given point propagate to other points in space at the velocity c. In one second, therefore, the distance covered is 300 000 km. Along the whole of this path there would be alternating increases and decreases in the strength of the fields, corresponding to 100 million undulations. Between each maximum there is therefore a distance equal to 300 000 km divided by 100 million, i.e., 3 m. Three meters is the wavelength of electromagnetic waves that oscillate 100 million times per second. This number of oscillations is called the frequency of the waves in question, which are given in hertz (Hz), in this case 100 megahertz (MHz). One hundred megahertz is a frequency used by frequency-modulated (FM) transmitters; in France their frequencies lie in a band between 87 and 107 MHz.

I will take this opportunity to describe the concept of bandwidth, which is crucial for SETI. Obviously, FM transmitters have to use different frequencies so that they do not interfere with one another. Imagine that there are 200 transmitters to accommodate in the band between 87 and 107 MHz; each will be able to use a bandwidth of 0.1 MHz or 100 kHz. With television, this is frequently called a 'channel'.

But why not have 1000 transmitters, each with a bandwidth of just 10 kHz? Here again, another fundamental concept intervenes, which we shall encounter frequently in our discussion of SETI. If a transmitter has an available bandwidth of 10 kHz, it is able to modulate its carrier wave at a range of frequencies between 0 and 10 kHz. It is the modulation that enables information to be imposed on the carrier wave. Without modulation, the waves would have a fixed frequency, and would be devoid of information, merely resembling a continuous whistling noise. If we want to transmit sounds (such as words or music), we have to use modulation, i.e., employ variations in the frequency or intensity of the waves that are emitted. However, that may only take place within the 10 kHz available in this particular case, so that it would be impossible to transmit any sound that was higher in pitch than the one that corresponds to 10 000 variations per second.

If that fails to satisfy music lovers, we must have fewer transmitters, because the bandwidth available sets the maximum frequency that may be transmitted. In fact, all this follows from information theory – the theory that describes the rate at which information may be transmitted – and from the theory of Fourier transforms. The latter are named after the distinguished French mathematician, Baron Joseph Fourier, who, in 1812, discovered trigonometric series, which are described by the Larousse dictionary very succinctly as: 'a mathematical technique of considerable importance.'

Decimetric radio waves

Although FM receivers are sensitive to electromagnetic waves with a wavelength of 3 m, our eyes are able to perceive waves between 0.4 and 0.8 microns, in the so-called visible range, between violet and red. Their frequencies are some million billion hertz. Ultraviolet, X-rays and gamma-rays have even higher frequencies, while infrared radiation at 1, 10 or 100 microns (the latter being 0.1 mm) form a transition to radio waves at millimeter wavelengths. A particularly important region for SETI is decimetric wavelengths, between 3 and 30 cm, with frequencies between 10 and 1 gigahertz (GHz).

Why should SETI try to detect decimetric wavelengths? It is a matter of what is practical. At the longest wavelengths, the sky suddenly becomes extremely 'luminous', because all sorts of celestial objects, including gaseous nebulae, radio galaxies, quasars, pulsars, supernovae, and supernova remnants are all natural and energetic transmitters at meter wavelengths and longer. Trying to detect emissions from a civilization at meter wavelengths would be like trying to photograph the stars in full daylight. The sky background would overpower the weak light from a star. This 'daylight' effect is obviously to be avoided.

Conversely, at wavelengths shorter than 1 cm, a deleterious effect, linked with the wave/particle duality of electromagnetic waves, comes into play. According to quantum physics, every wave is associated with particles (photons, in the current case), whose distribution and motion in space it governs. The greatest probability of finding the photons is where the wave is strongest. In addition, the energy of each photon is proportional to the frequency of its associated wave. As a result, for a given total energy transported by a wave, because of this quantum effect, the higher the frequency, the lower the number of associated photons.

Let us return to a civilization sending out signals. If it uses high-frequency waves, it will have, for any given energy budget, a relatively lower number of photons available for its transmission. Yet it is just these photons that carry the information. If we receive a photon, we say '1', and if we do not receive one, we say '0.' The sequences of 1s and 0s form the series of binary digits that encode the information. Consequently, if a civilization wishes to transmit the maximum amount of information (and thus of photons) for a given energy budget (the energy available at its transmitter), it should not use high-frequency waves.

To conclude this important question of the most favorable wavelengths for SETI, we find that the Big Bang plays an unexpected part. The background radio emission that is detected in every direction of the universe (the 2.7 K background radiation) is the fossil residue of the intense radiation that filled the universe just after the Big Bang. This background radiation is obviously a nuisance, and, although it is weak, it is unavoidable. When this is taken into account, it is found that the quantum effect that causes problems at relatively high frequencies does not exceed the level of the Big Bang effect except at frequencies above 30 GHz (with wavelengths below 1 cm). These three effects – the daylight, quantum, and Big-Bang effects – mean that the most favorable band for SETI is between 1 and 30 GHz.

The water vapor in our atmosphere prevents observations from the ground at around 20 GHz. Finally, then, if we add this 'water-vapor effect' – until we are able to install large radio telescopes in space – the best region for SETI is between 1 and 10 GHz, at decimetric wavelengths. Because radio astronomers are newcomers to this business of cosmic detection, and because they also tend to be pragmatic, they intend to start searching at these wavelengths.

Narrow-bandwidth signals

Having determined the most favorable frequency band at which to begin SETI – which is often called the SETI window – we need to examine the conditions under which a signal will propagate as far as possible. When radio astronomers – because they are the first SETI specialists to be involved – try to record a weak signal, their instruments are affected by all sorts of spurious signals: natural interference (e.g., from thunderstorms) and artificial forms (e.g., from industry); the random motions of electrons in the circuits of the ultrasensitive receivers that are used; the statistical fluctuations in the radio emissions from astrophysical sources, etc. All these more-or-less random fluctuations form what is called 'background noise.' We have to extract the signal from this noise.

In the case of a signal that is deliberately transmitted by a civilization in such a way that it travels as far as possible, and which might, therefore, be received at a very low strength (by us for example), it is best for the energy transmitted to be confined to as narrow a bandwidth as possible. Calculations show, in fact, that the signal will then be stronger for a given amount of energy, and that the interference from the background noise will be confined to that arising within the narrow bandwidth of the signal.

Consequently, if we are interested in searching for signals that are deliberately designed to carry information as far as possible, we need to concentrate on searching for signals with a very narrow bandwidth. This is, in fact, the American approach that NASA proposed to use in its SETI program, which has now been taken over by the SETI Institute. But other methods may be devised, in particular the Russian strategy, which is instead designed to detect stray leakage from astro-technologies, where the bandwidth would not necessarily be narrow.

Interstellar dispersion

The bandwidth of a deliberate signal may not be as narrow, however, as we might wish. In fact, during their journey, radio waves interact with the electrons that are found throughout interstellar space. Although they are few in number – about 100 per cm^3 – these electrons eventually have an effect on waves that take a long path through them. When a radio wave encounters an electron, the electromagnetic field forces the electron to oscillate, which causes the wave to change direction slightly, and to lose a little energy, so the wave's frequency decreases slightly. Over a path-length of ten light-years, waves of a few gigahertz, which when emitted all had the same wavelength, would be found on arrival to have been dispersed over a range amounting to about one-tenth of a hertz in width. This effect arises because of the random motions of the clouds in which the electrons are embedded.

This interstellar dispersion therefore means that it is pointless to transmit with a bandwidth less than 0.1 to 0.05 Hz, or to use receivers with bandwidths smaller than this range. The SETI receiver with the narrowest bandwidths is the Harvard one, where the bandwidth is 0.05 Hz. Note that there is a conflict between transmitting a signal over a long distance, and transmitting a great deal of information. The former requires narrow channels, and the latter wide channels. But that's the way physics works!

Without resorting to pure speculation, however, it is possible that a civilization might try to attract our attention by transmitting a signal with a very narrow bandwidth, whose only information would be to say 'Here we are'; and then use a wide-band, weaker transmission to provide us with detailed information. To detect such a civilization, we therefore need to employ more sophisticated techniques. Under these circumstances, the signals would have a composite bandwidth, consisting of a narrow, strong, peak frequency, flanked by a wide plateau of lesser intensity.

SETI's 100 billion channels

We can now consider the immense challenge that faces SETI: the SETI window (between 1 and 10 GHz) contains 100 000 million channels (each 0.1 Hz wide) for possible communication. It is worth recalling that 30 years ago, Frank Drake's first OZMA experiment (p. 113) was carried out with a single-channel receiver, and that the best receivers currently available to radio astronomers can handle 1000 channels simultaneously. One thousand compared with 100 billion! That is a factor of 100 million greater, to our disadvantage.

Although during the past 30 years a good half-dozen radio astronomical observatories have each managed to devote a few weeks of time to examining some hundreds of stars, all that they have managed to do is to cast their nets a few times into a vast ocean of interstellar signals, searching for a minute bottle that may, perhaps, contain a tiny piece of paper.

Salute the pioneers!

Nevertheless, we should pay homage to the pioneers, because, despite their knowledge of the desperate nature of their enterprise, they still went to great lengths to assemble state-of-the-art electronic systems, to develop analytic algorithms, and to overcome the ridicule of many of their colleagues, merely to find out, at the end of the day, that there was nothing in their nets, and that the authorities were disdainful, and not only refused to help them financially but also failed to give them moral support. Happily, many of the pioneers were made of sterner stuff; the elation engendered by working on such an extraordinary project protected them from becoming discouraged and from the resentment (often expressed in some very strange behaviour) of even experienced scientific colleagues.

Which observatories pioneered the search for radio signals? There were: Harvard University, working in conjunction with the Smithsonian Institution and Sagan's Planetary Society; the National Radio Observatory at Green Bank; the University of California, Berkeley; and the University of Ohio – all American. I take particular pleasure in also being able to mention the Paris Observatory at Meudon, where my colleague François Biraud, the CNRS Research Director has made, from the start, both scientific and technical contributions to the construction of the great radio telescope at Nançay. In 1970 he, together with J.-C. Ribes, now Director of the Lyon Observatory, published a popular, and extremely successful, book on SETI, *Le Dossier des civilizations extra-terrestres* [Dossier on extraterrestrial civilizations]. In eight different years since 1981, using the great radio telescope at Nançay, he has carried out 10-day programs, monitoring 300 of the closest, solar-type stars, including the first two studied by Drake, Tau Ceti and Epsilon Eridani. This was in direct collaboration with Jill Tarter (now at the SETI Institute), who subsequently became Scientific Director of NASA's SETI program, and who made special trips to work with him during his first, experimental listening sessions. This was how Biraud acquired the basic know-how that has enabled him to devise methods, strategies, and to construct electronic and computational prototypes. Above all, working on the job, soldering iron in one hand and keyboard under the other, he has been able to promote the idea of SETI in France, to plant the seed and help it survive for future enthusiasts. The French SETI tradition owes its origin to him.

The way is open. But should we be surprised that, despite the 100 000 hours of desultory fishing, no signal has yet been picked up? Should we abandon SETI and the wonderful idea of ever detecting an intelligent message from elsewhere? Luckily, once again, the frontiersmen from the New World, always ready to face the challenges offered by the exploration of new worlds, have picked up the gauntlet by developing new technologies.

8

NASA takes up the challenge

Exploring thousands of millions of decimetric radio channels represents a gigantic challenge. Nevertheless, 12 years ago NASA decided to take up that challenge. Yet again, it needed a revolutionary technology.

Ten million channels

To capture a particular channel, radio and television receivers use filters that reject everything except the chosen passband, and which are built of circuits consisting of self-induction coils and capacitors. For the last ten years, radio astronomers have been using filters based on auto-correlation circuits, the basic idea again being founded on Fourier's work (p. 128). If a wave with a very specific frequency is picked up by a receiver and, after being artificially delayed electronically by several periods, is superimposed on the wave received directly, the direct and retarded waves coincide. This electronic trickery, achieved by what is known as a delay line, therefore enables us to use the observed correlation to identify a wave of given frequency. The system acts as a frequency filter. This technology, commonly used nowadays, can provide 1000 simultaneous channels. Beyond that, it becomes cumbersome. Another technology has enabled some observatories to reach 10 000 channels, but it also runs out of steam beyond that level.

In fact, it was the arrival of large computers on the scene that allowed NASA to take the mathematical tool offered by Fourier's trigonometric series to the limit. If we feed the detected radio waves into a computer, we can subject them to Fourier transforms and look at the result. What we obtain is the intensity of the waves received at every frequency that we require: such-and-such a frequency, such-and-such an intensity. If, therefore, we 'listen' to a particular spot on the sky, and if a signal is being transmitted at a particular frequency, it will appear as excess power at that frequency, and produce a peak on the recorder. If the signal is being transmitted over a bandwidth of a number of hertz, it will produce a wider excess (a 'bump' on the recording), at the band in question. There are no limits to this technology: the mathematical operation offered by Fourier transforms is infinitely powerful. The only limitation is that imposed by the computer's capacity. This is the path that NASA decided to take.

The MCSA

Fourier transforms require intensive computation, the mathematical principles of which are simple but very specific. This is why NASA found it necessary to construct the necessary computer itself. The radio waves received by the radio telescope are amplified and converted to electric current by the receiver. They are then fed into a first set of 112 Fourier filters, each 74 kHz in width. The 112 currents obtained are then fed into 112 sets of 72 channels, each 1024 Hz wide. This gives 8064 currents, each of which is then subjected to further Fourier transforms, in parallel, by microprocessors, giving 1024 channels, each 1 Hz wide. By the end of this enormous cascade operation, we have obtained 8 257 536 individual electric currents which define the strength of the signal picked up by the radio telescope in narrow bands just 1 Hz wide. This extraordinary equipment is known by the rather barbarous abbreviation MCSA (Multi Channel Spectrum Analyzer).

Modular architecture

The chosen architecture, with successive stages, was preferred to a more conventional, monolithic architecture with eight million channels that could have been obtained using a Fourier transform superprocessor. It was, for example, easier to construct a more modest prototype (still with 74 000 channels, a record at that time), using a single subassembly of the elements involved. This highly modular architecture, with a large amount of parallel processing, reduces the number of different subassemblies, and localizes any failure to a single portion of the overall range of frequencies being analyzed. Finally, great flexibility is obtained by simply changing the microcode that specifies the successive stages of the transform. Above all, the system may be extended ... Its future is therefore wide open!

The prototype

The 74 000-channel prototype occupied an electronic rack, a vertical bay as high as a fridge-freezer. Flanked by its control and acquisition computer, like a stacked washing machine and tumble drier, the whole forms, in NASA's jargon, the 'electronic housekeeping' for any perfect, aspiring 'SETI household.' It was taken to Arecibo to 'wash' the signals picked up by the gigantic dish. After this test, NASA undertook the construction of the final MCSA. At this point, more advanced electronics were required. If the full equipment had been like the prototype, with cards like those in some of our old television sets, the full MCSA would have filled some 20 racks. Luckily, thanks to the modular architecture, there are only 13 different circuits among all those electronics; the main superprocessor contains 680 sets of these, all the same. So it is an excellent candidate for Very Large Scale Integration (VLSI), which enables a whole set of components occupying perhaps 10 cm × 10 cm by a single microchip with an area of 10 mm^2. The group developed a VLSI chip, which, reproduced 680 times, considerably reduced both the size and the costs.

How to envisage eight million channels

What would be the length of a hot dog capable of being cut into 8 257 536 slices each 1 mm thick? Quite obviously the length would be 8 km, 257 m and 536 mm. Laid out along the main axis of New York, it would reach from Battery Park to South Central Park. Every New Yorker could have a slice; if they queued up, with one every 50 cm, they would reach from New York to San Francisco. Serving one slice every 10 seconds, it would take 2 years, 7 months, 12 days, 4 hours, and 6 minutes to distribute them all, whereas MCSA has to 'swallow' all those slices every second! This may give some idea of the scale of the task that has to be tackled in searching for extraterrestrials.

The detector

If we wanted to visualize the strength of the signal received by all these millions of channels, we could use a pen recorder of the sort used for electrocardiograms; a pen would plot the values of each channel along a strip of paper, thus tracing out what is called a spectrum. If a channel or a small group of channels picked up a stronger signal than the neighboring channels, the trace would show a peak or a bump. All the rest of the spectrum would appear as a jittery line, with chaotic, irregular changes, more or less around the zero point, indicating the essentially random fluctuations in the strength of the signal caused by all sorts of variation in the transmission, the reception, or the amplification – 'noise' in electronics parlance. Even with one channel per millimeter, the strip of paper would still be nearly 1 km long – just for a single observed spectrum. It would therefore be unthinkable to try to examine it by eye.

This is why NASA also had to tackle the problem of the second part of the instrument, the signal detector. This is again based on mathematical techniques and electronic technology, and uses a specialized, sophisticated computer. This detector was designed and constructed to carry out a relatively simple task: recognize either a continuous signal, like a whistle, or a signal consisting of regular pulses, like 'beep, beep.' Even for two such simple types of signal, the task is still immense.

To see this, take the case envisaged by NASA. A candidate star is examined for a period of 1000 seconds; the spectrum coming from it is recorded every second. We obtain 1000 spectra, each containing 8 000 000 channels.

A nightmare image

Imagine that the computer uses a screen to display a spectrum as a horizontal line consisting of 8 000 000 consecutive points, the intensity of each point being altered to represent the strength of the signal received by the corresponding channel during that particular second. This is repeated on successive lines for each of the 1000 seconds making up the observation. What we have is a screen consisting of 1000 lines, each made up of 8 000 000 points, to give a total of 8 billion points (or pixels) of different intensities. By comparison, a television screen contains less than a million pixels.

If the luminosity is also encoded, the observation gives us a screen containing 100 billion units of information (or bits). From this plethora of information, we have to decide, immediately, whether there is, or is not, a signal, and do this in 30 minutes, before the next observation. This is because it is absolutely out of the question to store these 100 billion bits on magnetic tape or optical disk and analyze them later; that would serve only to postpone the problem.

In addition, if an observation of a star reveals a signal, we need to be advised immediately, so that we can begin studying it as soon as possible, to avoid running the risk that it will cease, for whatever reason. It is also essential to realize that, because of noise, there are innumerable pixels, some bright and some dark, scattered at random all over our gigantic, nightmarish display. (You can see a miniature version on your TV screen when the transmitter shuts down, and all you can see is 'snow', which is also caused by all sorts of fluctuations.) Given that NASA was to search for signals that are probably very weak – because of the colossal distances involved – a whistle or a 'beep, beep' would be completely drowned.

Extracting the signals

A simple 'beep, beep' signal, for example, consists of a series of regularly spaced points on the screen. If the pulses have a period of 10 seconds, 100 points will be lost among the 8 billion on the screen. In addition, there will be a shift between one point and the next, amounting to several channels, because of the inherent variations in velocity producing a Doppler effect during the observation. (The terrestrial receiver is moving with the Earth, and the extraterrestrial transmitter is also in motion.) The signal detector therefore has to search for an unknown number of rather weak points, spaced at unknown intervals along an unknown curve. It needs to carry out various calculations, which are relatively simple from a mathematical point of view, but which are horrendous when it comes to the number of combinations that they need to test to give a definitive answer in 30 minutes. Algorithms have been developed that are able to detect appropriate patterns. To reduce the amount of memory required to carry out the calculations, methods of elimination are employed. Several hundred microprocessors are required, each handling 100-kHz sections of the spectrum. Once again, VLSI is essential; the highly specialized processors, based on content-addressable memory, each contain 50 VLSI chips.

Thanks to these important technological developments, it has been possible to achieve 10 million simultaneous channels. Thoughts are beginning to turn to 100 million, and even a billion channels. When Frank Drake gave a summary at the end of the symposium at Val Cenis in 1990, he stressed the fact that, since his first attempt, not only had the number of channels risen from just 1 to as many as 10 million (and soon to 100 million) but the sensitivity of receivers had also risen by a factor of 1000. These advances have followed an amazing, exponential curve and point to an incredible future. The barriers are likely to be broken down, and SETI looks certain to succeed.

So why not wait a while before starting SETI on a large scale? This is a pointless question, because if Drake, with his single-channel OZMA project 30 years ago had not been followed by others who took up the hunt with 100, 1000, then 100 000 channels, technological progress would have remained a mere pious hope, and MCSA, together with its signal detector, would not exist. Since 1992 it has been possible to carry out in a single minute as much searching as would have taken OZMA 100 000 years. It is this incredible figure that has finally enabled us to reach the point where we can begin to search the notorious 'cosmic haystack' for the proverbial needle with some chance of success. We have to search for the right signal, from the right star, with the right channel, and at the right moment.

The listening program

The right signal, the right star, the right channel, and the right moment. That says everything! Our four-dimensional cosmic haystack is, in fact defined by those four factors, and its gigantic size causes us to pale into insignificance. We have to find the 'right channel' among 100 billion; the 'right star' among 100 billion for our Galaxy alone; and as for the 'right moment', our equipment must be operating when the waves of any possible signal actually reach the Earth. In an all-sky program taking several years, the listening time will last only a minute. Finally, the 'right signal' must be present, because, despite the enormous effort that is involved, with current technology the only signals for which we can scan are a whistle or a 'beep, beep', whereas far more complex signals are conceivable and have, as yet, attracted theoretical attention from just a few researchers. But radio astronomers are pragmatic, rational individuals, not lacking in courage, so they have buckled down to the task, encouraged by the prospect of advances in technology and substantial financing – at least in the USA.

What are the chances of success with the eight-million channel system, which was symbolically inaugurated on the 500th anniversary of Christopher Columbus' great discovery? Pragmatically, and with Olympian calm, I would reply: 'We will find a signal next week, or perhaps in a century ...' But if one *is* received, and that ought to happen, it would be fantastic. In subsequent years, there would be more, many more, from dozens of different civilizations, because once we have succeeded, we would never stop searching. Remember what happened with quasars, pulsars, gravitational lenses, amino acids in meteorites, etc. Once the first discovery has been made, many others follow.

The targeted search

NASA's listening program had two facets. The first consisted of monitoring 1000 target stars. This was the official program, the one that scientists used as a persuasive argument to obtain the necessary finance to carry out the program. To convince the authorities – the White House, the Senate, and the House of Representatives –

who are all specifically involved in passing any project requiring a significant amount of finance, the radio astronomers based their arguments on facts: specifically on the existence of life, intelligence, and civilization on Earth. A unique instance, unfortunately, but one that is unanswerable. We may recall our three reasonable working hypotheses: life on Earth is a result of the natural evolution of the universe; what happened on Earth could have happened elsewhere; and human intelligence is unlikely to be the highest form to have evolved in the universe.

One thousand target stars

Basing the initial steps on the facts of our own existence, it was decided to monitor the closest stars (to have as strong a signal as possible), that most closely resemble the Sun (to increase the chances of finding a planet similar to the Earth). This gave rise to the program of target stars, intended to observe 1000 stars, out to a distance of 100 light-years.

I must emphasize how extremely modest this program was, on two counts. First, such an investigation, out to a distance of 100 light-years covers a derisory volume of space, because our Galaxy is a vast aggregate of stars, some 100 000 light-years in diameter, containing more than 100 billion stars. We are exploring just one-thousandth of its diameter, and only one-hundred-millionth of its volume! Second, despite the fact that a targeted program will last ten years, the time spent on listening to any particular star on any given 1-Hz channel is – half a minute! If, during the ten years, a civilization broadcasts continuously, its signal will be captured – provided it is strong enough – during that fateful 30 seconds. But if the star emits radiation, even continuously, in a concentrated beam, like that of a powerful radar with a beam-width of 1°, for example, the beam will sweep the Earth only when it happens to be pointed in our direction. The chance of looking at the star at the correct moment drops by a factor of 40 000 (the number of beams 1° across required to cover the celestial sphere). For the 1000 stars on the ten-year program, therefore, the chance of coming across the right star, in the right channel, at the right moment, will be 1 in 50.

NASA's vast program was merely a beginning, happily sustained by some highly promising technological developments.

One million accessible stars

I have said that a signal will be detected if it is strong enough. Considerable efforts have been made in this respect; by using particularly sensitive receivers and radio telescopes with large surface areas to increase the quantity of radiation collected. It is estimated that if other civilizations have at their disposal methods comparable to our own, such as planetary radars like that at Arecibo, or radars similar to the military ones that maintain surveillance over the Arctic, we would be able to detect them out to a distance of 1000 light-years, i.e., ten times the limit chosen for the ten-year program, which implies a volume of space 1000 times greater. One million solar-type stars would therefore be accessible, as far as broadcast power is concerned, provided our systems become faster. If we do not want to spend 10 000 years listening to these one million stars, we need to increase the number of simultaneous channels, and aim for ten billion, which would seem to be feasible in a few years, with a few additional tens of millions of dollars.

Is the program too anthropomorphic?

Frequently, when I give lectures, I am reproached for the fact that the target program is too anthropomorphic, based on far too restricted a view of intelligence in the universe, namely on our own. In fact, although we took this as our starting point in scientifically justifying a program, we are very conscious that advanced intelligences or technologies may not require a terrestrial-type planet on which to emerge, nor even be based on carbon chemistry. As soon as we embark on this question, however, we have to contend with our utter absence of any information on this point, which prevents us from making much progress in justifying a more expensive program. Quite frankly, we fall rapidly into pure speculation or science fiction. These have never been noted for opening the financial coffers.

Nevertheless, Carl Sagan and Ed Salpeter, who are colleagues at Cornell University, one in the Planetary Studies Laboratory, and the other in the Nuclear Physics Laboratory, have amused them-selves by using their knowledge of science to imagine two possible forms of creature that could live in the atmosphere of Jupiter. The lower levels of this atmosphere are extremely warm and the outer-most ones are cold, but there is a layer where it is a comfortable 20 °C and where organic chemistry could evolve. The two scientists imagined living balloons, with organic envelopes that are able, by means of a mechanism for separating different gases, to float at this ideal level. To quote Sagan in his book *Cosmos*: 'A floater might eat preformed organic molecules, or make its own from sunlight and air ... the bigger a floater is, the more efficient it will be. Salpeter and I imagined floaters kilometers across, enormously larger than the greatest whale that ever was, beings the size of cities.' Such speculations forbid us from asserting too dogmatically that life is impossible on Jupiter, but they cannot justify an expensive veri-fication program. No one would risk a million dollars on such a prospect.

The sky survey program

If an instrument that has already been financed can be adapted, at minimal cost, to investigate a relatively speculative, but extremely interesting possibility, scientists are duty bound to make the at-tempt. This was the rationale behind NASA's sky survey – now, unfortunately, terminated by Congress. Given that the MCSA had been built, why not make a copy, or a variant, and survey the whole sky, trying to pick up any interstellar radio beacons of unknown na-ture? The word 'beacon' can stand for any civilization that is not, perforce, based either on a planet or on macromolecular life, but which is one capable of producing, intentionally or unintentionally, powerful radio emissions. This again evokes the idea of another science fiction speculation, Fred Hoyle's 'Black Cloud', wandering in interstellar space and manipulating substantial amounts of en-ergy. To sweep the whole sky, again over a decade, the program would have used medium-sized radio telescopes ranging from 30 to 40 m in diameter, to ensure that a sufficiently large area of the sky was covered at any one time. In addition, its sensitivity would have been less than the targeted search and with broader channels, again to save time. Obviously wanting to cover the whole sky is far more ambitious than just targeting 1000 stars, so we had to accept various compromises concerning the resolution of the scans, the width of the channels, and the sensitivity of the equipment.

The SETI system

Although the MCSA and the signal detector are the vital components of a SETI system, there are other significant elements. First, we need a radio telescope to capture the signals. Generally, such a telescope consists of a large bowl of metal mesh, which concentrates the waves at a point known as the focus, just as a concave mirror does for light. For decimetric radio waves, a mesh is just as efficient as the polished surface of an optical mirror, provided the sizes of the apertures and surface irregularities are less than one-tenth of the wavelength. A concave shaving mirror may be used to concentrate the Sun's rays to a small, bright patch, which may be sufficiently intense to ignite a piece of paper. The bowl of a radio telescope serves the same function for radio waves, concentrating them at a focus where the equipment for picking up the radiation is located. In principle, this equipment may be no more than a simple dipole (a pair of parallel rods or wires) like an ordinary television antenna. Other, more efficient, arrangements have been developed. With the Arecibo telescope, the system at the focus is a long 'pencil' carrying a series of dipoles at specific, precisely determined, spacings. At Nançay, the waves are captured by a sheet-metal, rectangular-section 'horn.' At the bottom, a parabolic section of the wall concentrates the waves onto a waveguide (which is effectively a tube with a carefully calculated section) that leads them to an electric inductance, in which the waves produce weak electric currents, which are immediately amplified by sensitive electronics. Only subsequently are these amplified currents fed to the MCSA for the Fourier-transform frequency analysis. Finally, the results of all the calculations are examined by the signal detector to see whether a whistle or a 'beep, beep' is present.

Even then the work has not finished: the detector's response is not inevitably a firm 'Yes' or 'No.' When we are searching for very faint signals, fluctuations in the background noise may accidentally resemble an artificial signal. It is therefore necessary to measure the background noise continuously to know how far we may go without being plagued by these random signals, but without unnecessarily restricting the efficiency of the search.

We expect to pick up stray signals, so the results from the detector are compared with a record of such (more or less identified) signals. A number of our own devices, such as radars, planes, and satellites, also produce artificial signals. To distinguish these, their specific characteristics, such as position and frequency must be determined. An extraterrestrial civilization, for example, would have a fixed position relative to the stars, whereas a satellite moves. When an artificial signal is detected, it is essential to check that it does really come from the target star, and that it is not entering the radio telescope from one side. To do so, the axis of the telescope is shifted slightly to one side; if the signal disappears, it was coming from the star. The frequency also serves to distinguish the source of the signal; the Doppler shift produces characteristic frequency shifts, for example, in the signal from a plane that is approaching or moving away from the radio telescope.

Even more complex systems are being developed to try to avoid the future increase in stray signals caused by our own civilization, for example, by linking the radio telescope to another in the vicinity. The Doppler effects will be slightly different for each of them, and analysis of this slight difference should enable us to reject signals produced by our own devices.

A SETI system is, by definition, extremely complex and requires an administrative computer to select the zones to be surveyed, to choose the frequencies, carry out verification tests, and, finally, to alert the duty astronomer if a valid signal is found. When this happens, the program will be changed immediately to concentrate on the possible candidate, whilst alerting other radio astronomy institutes that are able to carry out SETI observations. We dare not lose any time, because a true signal might stop at any moment. It is therefore essential to carry out immediately all the tests that are within our power to ascertain that it is a proper signal, before making any ill-considered announcement. To do so, we need to follow the source continuously, regardless of the Earth's rotation, by using a series of specialized radio telescopes spread over various continents. It is because of these eventualities that the SETI Committee of the International Academy of Astronautics has published an Announcement Protocol, and why I had originally suggested setting up a global SETI network.

Project Phoenix

Slightly less than a year after NASA's SETI program was inaugurated, Congress voted to terminate all the project's funding immediately. This startling decision not only brought NASA's targeted search and sky survey progams to a sudden end, but also affected Project META (see p. 163), the University of California's SERENDIP III project (see p. 165), and Ohio State University's program (because all three received some support from NASA grants), as well as the prospect of holding an international conference on SETI and Society at Chamonix Mont-Blanc in France.

During its short period of operation, NASA's targeted search monitored 24 solar-type stars for a total of 200 hours, using the Arecibo radio telescope. The sky survey with the 34-m antenna at Goldstone in California lasted for 1000 hours, and covered a total of 2646 square degrees or about one-sixteenth of the whole sky. Unfortunately the second program was closely linked to the antennas of NASA's Deep Space Network, which means that it cannot be undertaken directly by any other organization.

The prospect is much brighter for a targeted search. Reborn as Project Phoenix, this is being undertaken by the SETI Institute in Mountain View, California, whose President is Frank Drake, and which relies on private funding. Initial plans are for the equipment from Arecibo to be upgraded to 14 million channels, and installed on the 64-m antenna at Parkes in New South Wales, Australia. After a six-month period during which it will scan the southern sky, the equipment will be moved north again, initially to Arecibo, and possibly to other radio telescopes, such as the French radio telescope at Nançay, at a later date.

Although funding remains the major problem, the SETI Institute is already considering plans for the future construction of dedicated antennas in the 100-m class. Such antennas could, of course, be used for a revived sky survey program.

9

False alarms

Epsilon Eridani

Drake experienced the first of SETI's false alarms during his OZMA
project. Let him put it in his own words:

> Then we turned the telescope to point at our second target, the
> solar-type star Epsilon Eridani ... A few minutes went by. And
> then it happened. WHAM! We heard bursts of noise coming out
> of the loudspeaker eight times a second, and the chart recorder
> started banging against its pin eight times a second. We had
> never seen anything like this before in all the previous observing
> at Green Bank. We all looked at each other wide-eyed. Could
> it be this easy? ... Suddenly I realized that there had been a
> flaw in our planning. We had thought the detection of a signal
> so unlikely that we had never planned what to do if a clear signal
> was actually received. Almost simultaneously everyone in the
> room asked 'What do we do now?' Change the frequency? Well
> the most likely source of a spurious signal was the earth, and we
> could check that by moving the telescope off the star and seeing
> if the signal went away. So we proceeded to do that, and as we
> moved off the star, sure enough the signal went away. So we
> pointed back at the star. The signal did not come back. Wow
> ... There was all that adrenaline flowing and no way to apply all
> that excitement and energy in a useful way ...
>
> Day after day ... we turned to Epsilon Eridani ... A week
> went by and the signal didn't return. To our chagrin, one of
> our employees called up a friend in Ohio and told him what
> had happened. The word was passed to a newspaper reporter
> friend, and suddenly we were deluged with inquiries about the

mysterious signal. 'Had we really detected another?' 'No.' 'But you *have* received a strong signal with your equipment?' 'We can't comment on that.' And so, aha, we were hiding something. To this day many people believe falsely that we received signals from another world, and that some fiendish government agency has required us to keep this a deep dark secret.

This memorable alert happened on 8 April 1980. Ten days later, the signal returned and Drake eventually found it to be emitted by a high-flying passing airplane, perhaps a U2 stratospheric spy plane.

Little Green Men 1

Eight years later there was a new alert among English researchers quite unconnected with SETI. This time it was the unexpected discovery of pulsars as a result of their regular signals. Cautiously, the scientists wanted to analyze the matter before announcing anything at all. Among themselves, they called the source 'Little Green Men 1' (LGM1); then they discovered LGM2, and later LGM3. That was too much! They could not be artificial. Eventually, they discovered the true nature of the sources: rapidly rotating neutron stars – the famous pulsars. Such examples show that we need to proceed cautiously; the scientists involved need to have the chance to analyze the situation in peace. To provide evidence in support of this point, I made a historical and scientific study of one of the most unfortunate alerts, the CTA 102 affair.

The CTA 102 affair

In 1963, shortly after the pioneering work by Drake, Morrison and Cocconi, Kardashev, at the Sternberg Institute in Moscow, published a fundamental paper in the *Astronomicheskii Zhurnal* about 'The communication of information by civilizations on other worlds.' He discussed in detail under what conditions the maximum amount of information might be transferred, and maintained a different position from that later adopted by NASA, where preference is given to the detection of the simple, 'here we are', type of signal. In addition to his ideas about civilizations of Types I, II, and

III, Kardashev set out the characteristics of transmissions designed to transfer the maximum amount of information. These are a very broad spectrum, with a maximum in the decimetric region, variability on very short time-scales (thus indicating their artificial nature), as well as other properties. From catalogs of radio sources, he selected two promising candidates: CTA 21 and CTA 102. (These are numbers 21 and 102 in the first, *A*, catalog issued by the *California Institute of Technology*.) The stage was set for the events that were to unfold.

At the time, astronomy was beginning to come to terms with what was then a new, unknown field of research. In 1960, the radio source 3C 48 (the 48th in the *Third Cambridge Catalogue*) was identified as a 'quasi-stellar object' – in principle, it was regarded as being a star. In March 1963 in the United States, the first measurement of the recession velocity of a quasi-stellar object, 3C 273, was made. This revealed the fantastic value of 50 000 km/second, placing the object far out in extragalactic space. This was the first 'quasar' to be discovered; 100 times as bright, intrinsically, as a galaxy, and no longer a simple, individual star. During this period, Soviet astronomers, who were more or less up to date, and influenced by Kardashev's ideas, observed the two CTA sources with the Pulkovo radio telescope. They confirmed the very broad shape of the spectra!

In another approach, Sholomitskii, Kardashev's departmental head at the Sternberg Institute, assiduously observed the sources between August 1964 and February 1965 and found that CTA 102 varied in intensity by up to 30 %, and that the fluctuations occurred very regularly, with a period of 100 days. Such behavior was entirely consistent with Kardashev's suggestions. He also came to the conclusion that the body was within our Galaxy, and thus relatively close. In the light of these results, Sholomitskii issued a press release on 12 April 1965, through the Tass news agency, announcing that Soviet astronomers had observed signals that could come from extraterrestrial intelligence. On 14 April, he gave a press conference in Moscow about CTA 102.

In the meantime, in November 1964, two Americans had identified the radio source CTA 102 as being a quasi-stellar object, and on 8 April 1965, the Dutch-American astronomer Maarten Schmidt, at Mount Palomar, revealed this object's enormous recession velocity, thus putting it in the category of a very distant quasar. It is disconcerting to find a Russian announcing that the emission from an object was artificial, just six days after an American had identified it as a quasar. It should be said that, at the time, contacts between the two sides were not easy, even though there were no actual restrictions on scientists. Visas were difficult to obtain, telephone communication was not readily available, and fax machines were not widely used.

Sholomitskii carried on monitoring CTA 102 until November 1965, and continued to observe the same remarkable variations, whereas other radio astronomers around the world did not observe anything comparable. As one of them said, at the end of a report written in 1967, 'None of the others used precisely the same frequency or polarization', which was a cautious reservation by a prudent scientist, leaving room for uncertainty. Some twenty years later, however, CTA 102 is regarded as a perfectly typical quasar, one of the 6225 quasars in the 1991 catalog.

Following my investigation, in which I made use of the original publications, I proposed that the SETI Committee of the International Academy of Astronautics (IAA) should strongly support a resolution stating that 'before undertaking any public action in a SETI matter, it should be confidentially reviewed by an international, interdisciplinary committee of scientists, acting under the aegis of several of the international scientific unions.'

My own false alarm

By way of introduction, let me quote the description I wrote originally elsewhere:

> It is early morning, 18 March. Dawn is struggling to break on a depressing, cold, wet and windy morning in a gloomy Bavarian suburb. A group of us, from all parts of the world, are standing in front of the Tourotel, an apparently randomly chosen gaggle of human beings. Some have thick, comfortable coats, others have jackets, and some of us, shivering on the pavement, are even wearing short-sleeved shirts: 'yes, I have arrived before my suitcase!'

I manage to squeeze my way through to Geoffrey: 'Hi! Did you get my letter last week?'

'No, I had already left ... '

Here is the bus at last, spraying muddy water over us, and pouring out dirty exhaust fumes. At least we will be warm, and will be in time to hear the latest about the cosmic background radiation. Gently swaying to the motion of the bus, Geoffrey continues: 'What was your letter about, Jean?'

I move up closer, and whisper in his ear: 'Well, I've found abnormal radio emission coming from the direction of a solar-type star ... '

I didn't have time to say any more; immediately, Geoffrey turns to me, fixing me with his piercing eyes, and bursting with delight and hope, shouts: 'YOU'VE MADE CONTACT?'

Was it not an extraordinary moment? It happened to me in March 1986, when I was with Geoffrey Burbidge, one of the most clear-sighted and forward-looking of all astrophysicists, who, with his wife Margaret, was one of the pioneers of quasar research. With Fred Hoyle, he was a supporter of the Steady-State Theory, the rival to the Big Bang, and a defender of Halton 'Chip' Arp, who wholeheartedly pursued anomalous spectral redshifts, and who was banned from Mount Palomar by the powers-that-be. We were going to a symposium entitled 'Cosmology, astronomy and fundamental physics' being held at Garching, a northern suburb of Munich.

Originally, I was not particularly preoccupied by SETI; for ten years I had been observing galaxies with the brand new, large radio telescope at Nançay, as part of a small team that I had assembled, all involved with study of the 21-cm line. Then, at a meeting about supernovae in a small town in the south of Italy, I met an astronomer from the Brera Observatory, near Milan, Caterina Cassini. I was to be associated with her for a number of years, because we had both discovered what are now known as 'clumpy galaxies.' Combining our experience, hers at optical wavelengths, and mine at radio ones, we investigated them with some of the most powerful and newest instruments in the world, ranging from the largest, the giant Soviet 6-m telescope in the Caucasus, to some of the smallest, in Central Asia. At Alma-Ata, in Kazakhstan, one of these small telescopes

was the only one able to obtain the spectra of our galaxies, thanks
to its brilliant maker, Eddik Denijuk, who, at a glance, and using
merely an old microscope, could sort out intensities and velocities,
a task now undertaken by powerful computers.

It was during these wanderings that, in 1980, using the giant
radio interferometer at Westerbork, in the north of the Netherlands,
Caterina and I obtained a radio map of one of these clumpy galaxies.
In one corner, a splendid, compact radio source coincided with a
faint, reddish stellar image. Was it a star, or a quasar? Strange.
But for a year, I thought no more about it.

If the object was a quasar, however, it would be one of the bright-
est known, and it was red, rather than being more or less blue. So,
in 1982, when I was observing in Japan, I asked my colleague Sin-
Ichi Tamura to take a spectrum, just to see. It was incredible: a
spectrum of a solar-type star!

I became very excited. If it was such a star, it was hundreds
of light-years away. Its intrinsic radio emission was therefore extremely powerful, which was quite amazing for a star. Could it be
artificial emission? I needed to check this as soon as possible!

Luckily, the year before, I had observed the same galaxy with
the Very Large Array (VLA) in New Mexico, the American su-
perversion of the Westerbork telescope. The map obtained by my
colleague Dave Heeschen, the VLA's founder, did not cover the object. So, in 1983, I asked him if, by rerunning the tapes, he could
extend the map. Once again, the object was there! Its emission had
lasted for at least a year.

Victory? No, not yet. There was a tiny shadow on the horizon,
a tiny technical detail. Because the object was just on the edge
of the VLA's field, its radio image suffered from the equivalent of
the chromatic aberration that occurs at optical wavelengths; that
implied that the emission was taking place over a broad frequency
band.

Perhaps, after all, it was simply one of the rare stars active at radio
wavelengths. But, when I had checked the archive of photographic
plates obtained with the Japanese wide-field telescope, at Kiso, I had
been able to estimate the star's distance from its color. Its intrinsic
radio emission would have to be a million times greater than that
of the Sun, which was an enormous value! Scouring through the

scientific literature, I learned that, from time to time, a few, rare stars occasionally exhibit radio bursts that reach levels 100 times greater than normal, and for about one hour. I was still far short of a factor of one million! I was stuck; I needed more observations. Might it not simply be a perspective effect? Perhaps, by chance, a solar-type star happened to lie directly on the line of sight to a very distant radio source ... This possibility did not stop me from being preoccupied by the problem. To myself, I wondered if a civilization based on that star had used its asteroids to build a vast radio reflector, 100 million km in diameter, to focus the ordinary radio emission from its star, and was sweeping the beam across the sky to attract attention ... An excruciating ethical question bothered me: just suppose that this daft, but physically possible, idea was actually the truth? I could no longer keep everything to myself. Suppose humanity lost the chance of making contact because of my negligence? I decided to obtain more information.

A few short, but properly targeted observations, made by well-placed colleagues, during one of their dead times (when their objects of study were not visible) would be sufficient. How should I alert them? Publish a note in a scientific journal? Out of the question, I did not have enough to form a serious basis for a paper. In a popular article? That would take a long time, and would run the risk of distortion. Through personal contact? This was my final choice. In February 1986, I wrote a confidential letter to 22 well-placed investigators, whom I knew to be favorable to the idea of SETI. Twelve replies brought some new material, but there was nothing that was decisive.

Then a miracle happened. In July, when visiting the splendid Romanesque church of Saint-Sernin, in Toulouse, I happened to meet Bernard Burke, the VLA's specialist in 'quick looks' at anything that, at first sight, appeared rather exotic, such as gravitational lensing effects. 'I need an extremely precise position', I said. 'Contact Cam Wade', he said, 'He's there at the moment.' Finally, in August, I had the missing piece of the puzzle: the radio source is very slightly displaced from the star, by an amount equal to the angle subtended by the thickness of a hair at a distance of 4 m.

My six-year alert was over. It was just a simple perspective effect. If the protocol that the SETI Committee is establishing had been in force at the time, the problem could have been solved much sooner, probably in less than a year.

Am I saddened by the fact that I did not discover an extraterrestrial message? I would have been the first, and I would have become as famous as Christopher Columbus. I would have been inundated with receptions and lectures, inaugurations and exhibitions, cocktail parties and decorations. I would been asked the most profound questions, and maybe even assassinated! Above all, however, I would have been happy to have brought off such a coup, and to have opened up a unique frontier: 'the ultimate "Final frontier"', as I like to say as a joke. The third millennium (according to Western decimal reckoning), might have opened with new hope. Mankind might have found a new solidarity, a common sense of purpose, when faced with others who were perhaps more evolved, or wiser ... Nevertheless, for me personally, I had experienced a quite fascinating false alarm.

10

The expansion of SETI

Although, because of the sophistication of its system and its level of finance, for some decades NASA was the leading organization involved in SETI, it was not the only one attempting to crack the problem of extraterrestrial intelligence. Apart from the pioneer efforts that have been discussed earlier, the first project that I want to mention is an unusual one, undertaken by a clear-sighted, tenacious researcher, who is working with financial support provided by contributions from the general public. The person concerned is Paul Horowitz, Professor of Physics at Harvard University, in Cambridge, just outside Boston, Massachusetts. To European visitors, Harvard's campus seems like a bridgehead of the Old World, established on the shores of the vast, new, American continent.

THE 'MAGIC' FREQUENCIES

Carl Sagan and Iosif Shklovskii

The catalyst and channel for contributions from the general public was the Planetary Society, founded by Carl Sagan. Apart from the courses that he gives and his research work at the Planetary Studies Laboratory at Cornell University, he has played a fundamental role in promoting the search for extraterrestrial life. In the purely scientific field, for example, he made major contributions to the Viking Landers. He has also developed organic chemistry simulations in an attempt to reproduce the dark, complex substances that are suspected to exist on certain celestial bodies. His experiments have produced substances to which he has given the generic name of tholins.

157

Above all, however, it has been with the general public, the various administrations, and, in fact, with all the decision-makers, from the White House to the Senate and the House of Representatives, that Carl Sagan's role as a publicist has been of fundamental importance for bioastronomy. His television series *Cosmos* became extremely popular, not solely because the subject of life elsewhere fascinates people, but also because of the unexpected points of view that he frequently adopted. The book of the series, full of striking illustrations, was a major success in the United States and elsewhere. The general public is avid to know, to learn, and to dream. Sagan has also written a novel, *Contact*, in which he describes, extremely realistically, the events surrounding detection of the first signal. (It is easy to recognize some of the leading figures in the SETI field among the characters.) He also collaborated with a Soviet astronomer, Iosif Shklovskii, on *Intelligent life in the universe*. This is an adaption of the latter's book, *Vselennaia, zhizn', razum* [Universe, life, and reason] and recounts the development of SETI in a simple, vivid manner.

Shklovskii, one of the foremost Soviet theoreticians, became famous well before the days of space probes, through an apparently ridiculous idea. According to him, the two small satellites of Mars were hollow, and thus artificial. He thought that they were hollow, because that appeared to be the only way of ensuring that their density was low, and a low density was required to explain a strange fact: slowly and inexorably, both martian satellites are approaching to the surface, into which they will crash in some ten million years. Shklovskii suggested that atmospheric friction, frequently encountered with our own satellites, might be the cause of their behavior.

Sagan may have had this in mind, when, one evening, after everyone else had finished examining the first photographs of Mars sent back by the probes, and had gone to bed, he concentrated on the tapes that contained the first, apparently disappointing, raw data of a martian satellite. There was no clearly visible image, but, by repeated computer programing and image enhancement, by dawn he eventually obtained the first real image of another planet's satellite. Instead of resembling a giant, artificial, hollow space station, it looked more like a lumpy potato. This image is well known nowadays, and in 1992 the first asteroid ever to be photographed, Gaspra, proved to be another example of a similar-looking object.

The Planetary Society, founded in 1980, has 100 000 members. The subscriptions help to finance various projects aimed at exploring the universe, particularly those that are either insufficiently funded, or have been neglected by government sources. Among these scientifically valuable projects we may mention the study of balloons for exploring Mars that is being carried out by France's CNES, and Paul Horowitz's, Harvard-based, SETI program. The later carries the American logical argument to the extreme: use the narrowest possible bandwidth for listening. This minimum width is set by the scattering caused by electrons in interstellar space, and amounts to around one-tenth to one-twentieth of a hertz. In comparison, NASA's program (and its successor, Project Phoenix) was not intended to use bandwidths below 1 Hz. With its ten million channels, this means that it covers an overall bandwidth of 10 MHz. Even with this overall bandwidth, 1000 different ranges are required to cover the whole of the SETI window. This explains the length of NASA's program, which was expected to run for ten years.

Although it also has ten million channels, the Harvard equipment covers only 500 kHz, and it would take two centuries to cover the whole SETI window. This again is a measure of the enormity of the task facing us. We can thus understand the necessity for compromise, given our current state of technology. Because exploration of the whole of the window is out of the question with this system, use is made of the concept of 'magic' frequencies.

The 21-cm line

The radiation emitted naturally by neutral hydrogen atoms, the most abundant atoms in the universe, occurs within the SETI window (p. 129). This frequency is 1.420405751786 MHz, or a wavelength of 21.106 cm. This is the famous 21-cm line. This line arises from the interaction of the spin – the moment of angular rotation, in quantum mechanics – of the single electron that orbits the atomic nucleus (which itself consists of a single proton), with the spin of the proton. This interaction may be likened to that between two small, parallel, bar magnets. When the two magnets are anti-parallel, i.e., with opposite north–south orientations, they attract one another, whereas when they are parallel, with identical north–south orientations, they repel one another.

The energy states differ depending on whether the spins of the hydrogen atom's electron and proton are parallel or anti-parallel. Quantum physics allows us to calculate this energy difference, which is very small. Normally, the hydrogen atoms adopt the most stable state, corresponding to the lowest energy level. If they are disturbed, however, they may jump to the higher-energy excited state, and remain there (in the calm of interstellar space) for ten million years before spontaneously reverting to the lower state.

When an atom undergoes this transition, it emits a photon with a wavelength of 21 cm, which corresponds to the difference in energy between the two states. This is the origin of the 21-cm radiation. It arises from the atoms of neutral hydrogen that are found throughout interstellar space, where they are undisturbed for sufficient time for them to emit their radiation spontaneously. This is why the 21-cm radiation is the most widespread natural phenomenon in the universe. Its frequency is a magic number in that it is universal. This is what gave rise to the idea that, among the vast range of frequencies in the SETI radio window, it might serve as a common standard for interstellar communication. This is the basis of the strategy adopted at Harvard: explore in detail the range of frequencies around 1.420 GHz. Naturally, neglecting any signals that are not centered on the 21-cm line is somewhat risky. Taking such a risk does, however, offer other advantages. The extremely specific nature of such a narrow signal does ensure that there is a greater probability that it is artificial. In addition, because it is more easily detected among the noise caused by random fluctuations, it may be detected at a greater distance, with lower signal strengths.

The Doppler effect

This method does, in fact, offer an effective method of combating all sorts of interference. To appreciate the way in which this struggle – which is increasingly becoming of paramount importance – is being conducted, we need to refresh our memories about the Doppler effect. When you hear a distant ambulance siren in a town, you may notice that its pitch changes; it sometimes becomes higher or lower, and this may happen either suddenly or gradually. This can tell you something about the behavior of the vehicle even if you cannot see it. If the pitch decreases suddenly, the ambulance may have been coming toward you down a distant road, and then had to stop at traffic lights; or it may have been moving along a cross-street, and turned away from you at the next intersection. If the pitch rises and then drops, in a long glissando, the ambulance is probably circling a roundabout. In most countries, traffic moves counterclockwise, so the vehicle was moving from your right to your left. This indirect information derives from the fact that when a sound-source is approaching, the waves are bunched together, and thus have a shorter wavelength when they arrive at your ears, producing a higher note. When a source is moving away, conversely, the sound becomes lower.

The phenomenon, discovered by Christian Doppler, was found to apply to electromagnetic waves by the French physicist Hippolyte Fizeau. For the latter, the relative change in frequency is given by the ratio of the velocity of the source to the velocity of light. For example, if a planet approaching us at a velocity of 300 km/second emits a signal at 1 GHz, this would be received at a frequency of 1.001 GHz, or 0.999 GHz if the planet were receding at the same velocity. For a radio transmitter in an aircraft approaching at 300 m/second, the frequency would become 1.000001 GHz. It would be shifted by 1 kHz. For a vehicle moving at 30 m/second the shift would be 100 Hz, which is not very much.

The fight against interference

With the Harvard system, this would, however, correspond to a shift of 2000 channels each 0.05 Hz wide, which is enormous. Let us now see what happens if the speed of the vehicle changes, say by 1 %, during a typical observation lasting 20 seconds. The shift will affect a range of 20 channels around the 2000th channel away from the nominal frequency of 1 GHz that we are considering. The system makes use of this small difference to reject interfering signals, because, under the conditions that we have described, the interference signal from the vehicle is spread over 20 channels, so its strength is reduced to one-twentieth in each of the individual channels that are affected.

If the interference comes from a ground-based radar, for example, there will not be any shift. How then, can it be distinguished from an extraterrestrial signal? This is where Horowitz made use of the Doppler effect, which is extremely large with such narrow channels. The radio telescope is moving in space: because of the Earth's rotation (and depending on the latitude of the observing site) it may move at up to 460 m/second. In addition, the revolution of the Earth around the Sun causes it to move at 30 km/second, and the Sun carries it round the center of the Galaxy at 300 km/second. Because of these motions, an extraterrestrial signal will be shifted during the course of a typical observation in a complex, but calculable, manner, over hundreds of channels. All that is required, therefore, is to calculate, in advance, the appropriate compensation for the shift, to ensure that the signal falls in a single channel, where it will be reinforced. The radar interference, however, will be spread over a large number of channels, and thus drastically reduced.

Which reference system?

With eight million channels, each 0.05 Hz wide, the system covers a total of just 400 kHz, which, when the Doppler effect is taken into account, corresponds to a range of velocities of just a few tens of kilometers per second. This is enough for us not to have to worry about planetary velocities, whether our own, or of others, but is inadequate to cope with stellar velocities, which may amount to hundreds of kilometers per second. In adopting this strategy, therefore, we are also forced to make another assumption: it is not only essential in attempting to communicate that there is tacit agreement on the magic frequency, but there must also be agreement on the reference system employed. The transmitter and the receiver are both affected by various movements of astronomical origin, and therefore need to use the same reference frame.

The various forms of motion that occur in the Galaxy enable us to determine three 'magic' reference frames. The first is fixed by the general motion of the stars in our particular region of the Galaxy;

the second relates to the galactic center; and the third one makes use of the 2.7-K background radiation. Horowitz's observations are therefore carried out in these three reference frames, one after the other. The velocity corrections, and thus the frequencies, are applied by the computer.

Horowitz's system

Technically, the channels are also obtained by the use of Fourier transforms. To minimize the amount of electronic processing required, however, the degree of signal recognition has been reduced as much as possible: all it does is to indicate, at the end of every 20-second observation, which channels have received most energy. There is no way of distinguishing between a whistle and a 'beep, beep' signal. Nevertheless, it was still necessary to manufacture 20 000 integrated circuits, and make half a million soldered joints in constructing the system. The instrumentation was installed behind an old, disused, but refurbished radio telescope, 26 m in diameter, that belongs to Harvard University and the Smithsonian Institution, and which is sited in the mountains in Massachusetts. The radio telescope is aligned toward the south, but at a different elevation each day. Day by day, thanks to the Earth's rotation, it sweeps different parallels of celestial latitude. The whole sky visible from the site is thus covered every 200 days.

In five years of continuous operation, the system, dubbed META for *M*egachannel *Extra Terrestrial Assay*, has obtained eight million spectra, each consisting of eight million channels, and from which one megabit of data is extracted every month for the archives. When this 'treasure trove' is examined, 98 % of it is found to consist of noise, various forms of interference, operational errors, and 'other' signals – although not necessarily signals from 'others.' On examining the last category in detail, a few good, extraterrestrial candidates remain, one of which occurred on 10 October 1986 at 17:54 Greenwich Mean Time (GMT), but which was not repeated beyond the 20 seconds in which it was detected. All that can be done, with this alert, is to store it in the archives, and carry on searching ...

This is what Horowitz hopes to do, and he sets his sights high. As a result of technological advances since the beginning of the

1980s, 64 kilobit memories have been superseded by 64 megabit ones. Because the cost of the system is largely governed by that of the memory chips, it is possible, as he said at the Dresden SETI forum in 1990,

> We can expect ... within the next few years ... a gigachannel spectrum analyzer [BETA, or *B*illion channel *E*xtra*T*errestrial *A*ssay] for $150 000 ... In a university environment one can take advantage of 'slave labor' in the form of graduate students and eager undergraduates, resulting in a finished system for $250 000 ... These narrowband studies ... should be viewed as attempts to conduct meaningful university SETI exploration at low cost. Given the absence of successful detections, it seems entirely appropriate to field searches of greater power and comprehensiveness, for example the bold SETI NASA plans.

Can we make use of supernovae?

Another SETI strategy was suggested when the 1987 supernova in the Large Magellanic Cloud exploded. The idea is that when a civilization sees the explosion, it emits a signal in a certain direction, and this signal will arrive at the target star at the same time as it sees the explosion. The supernova therefore serves as a sort of starting pistol, coordinating transmission and reception, thus considerably increasing the chances of simultaneous activity. Coordination in time between transmitter and receiver applies to stars that are situated on the surface of a specific ellipsoid; as time passes after the explosion, the ellipsoid expands, reaching new stars, which may be identified by calculation. These are the ones that ought to transmit, at this precise moment, whereas we ought to choose them as targets for 'listening.'

The southern hemisphere becomes involved

The Argentinian Institute of Radio Astronomy has two 30-m radio telescopes, built to study southern-hemisphere galaxies at the 21-cm line. Its Director, Fernando Colomb, wanted to extend SETI to the southern hemisphere of the sky, most of which is invisible from farther north, so he started a listening program in 1986. After the explosion of supernova 1987A, the Argentinians directed their antennas toward the stars that fell on the ellipsoid. In addition, at the suggestion of an astronomer from Kharkov, they monitored the star that had intrigued me for so long. In his view, the radio source is slightly displaced from the star and could, therefore be coming from a transmitter located on one of its planets. Although the Argentinians did not detect anything, the exercise enabled them to get their hand in, and in 1986, with financial help from the Planetary Society, and technological help from Horowitz, they built a duplicate of his system. Since the end of 1990, this has been sweeping the southern sky, operating for 11 hours per day on one of their two radio telescopes.

Another search strategy

Attempts at SETI are gradually expanding and becoming established. An interesting possibility has been raised by workers from the University of California, Berkeley. One of SETI's great problems is that of obtaining radio telescope time for observations; competition is severe (although it is becoming less one-sided), with other astronomers interested in comets, stellar evolution, galaxies, quasars, etc. There is a solution: when an astronomer is observing, why not divert a fraction of the radiation that is entering the detector, and scan this for signals? This is what S. Bowyer and his collaborators at the University of California at Berkeley have done, in their SERENDIP project. Naturally, they have no choice over the point of sky that is being investigated, nor over the frequency employed, both of which are determined by the principal astronomer's program. When it comes to SETI, however, we have everything to gain, because there is no way of positively knowing which star or point in space might be a source of a signal, nor the frequency that may be used. There is all the more reason to try one's luck with free, and abundant, telescope time.

The diverter system operates autonomously and automatically, and originally used a 64 000-channel detector. The data are recorded on optical disk and examined at the university. The system operated for two years at the focus of the large 100-m radio telescope at

Green Bank in West Virginia, until the latter suddenly collapsed. The instrument did not disintegrate under the weight of the Berkeley system, however, but simply because of its age. It damaged the receiving laboratory sited underneath it, but luckily the observer on watch escaped injury. Nevertheless, during this series of observations, the team analyzed 300 billion spectral points, and extracted several million candidates. The latter have been subjected to stringent tests to eliminate most of the interference, leaving a list of promising candidates, for which specific telescope time will be requested for further observations. This very promising idea has led the Berkeley team to construct a receiver with four million channels (SERENDIP III) and hope to have a still more powerful one with 160 million channels by 1995, a new giant step indeed. In the face of such an onslaught, how can SETI fail?

After the demise of NASA's program in 1994, it is worth noting that SETI is, in fact, taking place all over the world. In addition to the adoption of the ex-NASA program by the SETI Institute and the continuing progress of the Harvard, Berkeley, and Argentinian programs, Ohio State University is expanding the hardware employed in its own search. This particular program is the longest-running search program in the world. (It was initiated in 1979!) Quite apart from these projects, programs are either in progress or are under development for large radio telescopes at Nançay in France, Parkes in Australia, Bologna in Italy, and Khodal in India.

Are there too many 'magic' frequencies?

The concept of a magic frequency, among the vast range of possible communication channels, at which cosmic civilizations would concentrate their efforts, goes back to Cocconi and Morrison and their first theoretical work in 1959. They accorded this particular status to the neutral-hydrogen frequency, 1420 MHz. Since then the principle has been extended to other preferred frequencies. The reasons for this may have been because none of the early projects detected any artificial signals at a frequency of 1420 MHz. However, given the extremely primitive nature of the first attempts, which had little chance of succeeding, there was not really any point in trying to find other possible frequencies.

Yet this is the way in which the Harvard instrument has been used. After having swept the whole sky at a frequency of 1420 MHz, without finding anything, it began again at double the frequency, 2840 MHz, which, according to Horowitz, may also be a significant value. Since resuming SETI investigations, which they had earlier abandoned, Australian radio astronomers have chosen a frequency of 1420 MHz times π (the universal mathematical constant that expresses the ratio between the circumference of a circle and its diameter), and are therefore using $1420 \times 3.1416 = 4462$ MHz.

Unfortunately, the more magic frequencies people think they have found, the less value can be accorded to each one. The basic concept becomes indistinct and lost. The worst of it is that with advances in radio astronomy, other notable emission lines have been discovered. One example is that of the hydroxyl radical, OH, second only to the 21-cm line in its strength and universal nature. This has a group of four lines around a wavelength of 18 cm. Ought we to listen at each of the four frequencies, or at the mean derived from the weights of their individual intensities?

Because OH and H combine to give water, H_2O, which is of fundamental importance for life on Earth, the band between 18 and 21 cm is regarded as a specially significant frequency region. In addition, it also falls within the SETI window, where fluctuations are at their smallest. It has therefore been called the 'Water Hole', indicating that it may be a meeting place for cosmic wanderers. Subsequently, lines of carbon monoxide, formaldehyde, water vapor, etc., have all been discovered. Finally, physicists have also discovered magic frequencies, quite independent of the universe's content of atoms or molecules. Just as Planck was able to calculate the famous Planck length (p. 5), 10^{-33} cm, by combining three of the fundamental physical constants, other combinations suggest additional special frequencies.

It is as well to note that, in this labyrinth, Ariadne's thread seems to be starting to unravel, and it is difficult to tell which way we should turn. This is why, NASA, setting its sights high, decided that *all* of the SETI window should be explored. 'All' in this context means from 1 to 10 GHz. Although this might not be a panacea for all our problems – remember that just 30 seconds will be spent on a particular star at one specific wavelength in ten years of scanning – it is an observational ambition of the first order, and one that we need to adopt, if we want to expand the field of SETI investigations. Obviously, to do so, we need to have means equal to the task.

A NEW ARIADNE'S THREAD

I have also succumbed to the attraction of special frequencies that may help us to find our bearings in this cosmic jungle. I do, however, adopt what was NASA's philosophy and now guides the SETI Institute: explore the whole window. Instead of covering the whole range from one end to the other, however, by successively shifting the set of ten million channels ten million times (and rejecting any form of compromise), I suggest monitoring a list of very specific, preferred frequencies, whose chances of success may be evaluated. I base this suggestion, not on physical constants, nor on the atomic or molecular properties of the universe, but on a specific class of exceptional objects.

This method offers us a way of plunging into the vast labyrinth exposed by NASA, but using a new thread, made from a different material, to guide us. If we find nothing at the end of the thread, then we will have to hand over to others, and let them systematically explore the whole field, as they intended to do originally. Even if nothing comes of it, at least this idea will not cause any loss of time: many frequencies will already have been explored. Because no one seems to be able to advance any arguments suggesting that this will lead us down a blind alley, there is increased hope of an early success in the long program envisaged by NASA. We might as well attempt to succeed in the first year rather than in the ninth.

Pulsars

The exceptional objects that I am proposing to use are the pulsars. Why? Because they are natural, interstellar radio beacons. They are powerful, well-distributed throughout interstellar space, emit pulses of staggering regularity, and have lifetimes of a million years. They are ideal radio 'lighthouses' that act as guides, not just in our own galactic neighborhood, but throughout the whole Galaxy, and even in external galaxies. The frequencies of their pulses are natural, stable standards, with long lifetimes, and they are visible at great distances.

From the physical point of view, a pulsar is a rapidly rotating neutron star. When a star arrives at the end of its evolution by exhausting its nuclear fuels, most of it collapses, and may produce an extremely dense remnant, consisting mainly of neutrons. For example, a typical star that is slightly more massive than the Sun, with a diameter of one million km, and rotates on its axis in about a month, ends up (after having ejected part of its mass in a supernova explosion), as a sphere 10 km in diameter that is rotating extremely rapidly.

Because of the intense electromagnetic phenomena that arise at the surface of this strange object, it emits two powerful beams of radio waves that sweep a whole region of space at every rotation. Each time such beams cross us, we observe a pulse, whence the name pulsar, for 'pulsating stellar object'. In the SETI context, however, remember that the first one was called LGM, Little Green Men!

Since their discovery, several hundred pulsars have been detected. In our galactic neighborhood one is found about every 500 light-years. Their rotation rates run from around 1 Hz to 1 kHz, corresponding to rotation periods of one second to one-thousandth of a second. They are so stable that over 1000 years their periods change by one-ten-thousandth of a second (at the most). In addition, the rates at which they slow down are known with sufficient accuracy for corrections to be made, if necessary. Would any cosmic navigator hesitate to set off in search of other civilizations with such lighthouses to mark the way?

When Carl Sagan and his wife devised a message to be carried by the Pioneer probes (the first human devices to leave the Solar System), and intended for any possible interstellar civilizations, they used pulsars. From the positions in space and the values of their rotation periods that were given, any beings intercepting the probes will be able to determine the position of the Sun, and the date when the probes were launched.

Targeted searching

To avoid becoming bogged down in technical detail, I will describe this strategy solely in terms of NASA's targeted program. The 1000 stars that it would have observed repeatedly over a period of ten years (and which will now be studied by the SETI Institute) lie within a sphere of 100 light-years radius, centered on the Sun. The pulsars closest to this region of interstellar space lie at distances of 260, 300, 490, and 550 light-years. Obviously the best beacon common to us and to all the 1000 stars is the closest: its astronomical name is PSR 1929+10 (PSR signifying radio pulsar; 1929, its Right Ascension, $19^h 29^m$; and +10 its Declination, $10°$ north). We should not, however, reject the pulsar at 300 light-years out of hand. If we need to set a value on their relative significance, we can easily assign weighting coefficients that are a function of the square of their distance.

Once the pulsars have been selected and an appropriate weighting devised, we face the major problem: their rotation frequencies are not related to the SETI window. What link is there between the two factors? We need to find a law that is as general and universal as possible. Bearing in mind that from one end of the SETI window to the other there is a factor of 10 (1 to 10 GHz), if we multiply the frequency of a selected pulsar enough times by a universal mathematical constant that is slightly less than 10, the final figure will, inevitably, fall in the SETI window.

For example, PSR 1929+10 has a rotational frequency of 4.4146768 Hz (note the accuracy; this is rarely achieved in astrophysics!). Multiplying this by 6.2831853 (2π, a striking, universal constant that is somewhat less than 10), 11 times in succession, we obtain a frequency of 2.65998 GHz, which falls within the window. To my mind, this is the first preferred frequency at which NASA should have started its targeted program in 1992.

Significant mathematical constants

But why choose 2π as a mathematical number? Could we not also use π; e = 2.7182818 (Euler's number, the base of natural, or Napierian logarithms), or even simply the number 2 itself, as recommended by Horowitz? What makes a mathematical constant significant? François Le Lionnais, who has been fascinated by numbers since infancy and is an extremely fine scholar, a champion chess player, and who was for a long time producer of *La Science en Marche* [Science Progress] on the French cultural radio channel, has collected these numbers all his life. In collaboration with Jean Brette, Director of the Mathematics Department of the Palais de la Découverte, he has published a book entitled *Les Nombres remarquables*, an anthology of 400 numbers and their 700 properties, ranging from 0.001264489 to $2^{86242}(2^{86243} - 1)$! You might think that there would be many between 2 and 6.2831853, but there are actually only 19!

In fact, we need not reject these numbers, but rather we should simply evaluate them as a function of the likelihood that they will bring success, which diminishes the more they differ from 10. On this basis, the frequency that scores most (100 points) is the one that we have already mentioned (2.65998 GHz); in second place is 2.38093 GHz with 79 points, two others with 62 points, etc. Overall, there are about 30, down to those gaining the rather low number of 8 points.

If we compare these 30 with the 100 billion possible channels of communication, we could say that this is an extremely powerful selection process: it is worth chancing one's arm on the basis of my list. In addition, this strategy based on pulsars appears to follow what, in our current state of knowledge, seems to be a universally applicable chain of logic. Finally, the frequencies obtained are very precise, to 100 kHz. This is particularly valuable for systems, like the one that Horowitz has devised, where the channels cover a total range of only 400 kHz. It enables the narrow bandwidth to be correctly positioned among the whole of the potential frequency spectrum.

The observational test

To add the finishing touches to the theory behind my arguments, I, together with my colleagues Jill Tarter and François Biraud, subjected this strategy based on pulsars to an observational test. We carried out a real search with the radio telescope at Nançay. In

memory of Frank Drake's pioneer research, we included his two
stars, Tau Ceti and Epsilon Eridani, among the targets. Unfortunately (or luckily?) they did not provide the surge of adrenaline
that he experienced during his first alert! We did, nevertheless,
have a valid alert: DM −23°8646, the 8646th star in the 23°-
south zone in the Bonner Durchmusterung Catalogue, compiled
in 1886. This star showed a strong signal at 1 661 590 123 Hz.
Three days later, we monitored it again, but the signal had disappeared.

Will this strategy based on pulsars make any converts? Woody
Sullivan, Professor of Astronomy at Washington State University,
an ardent promoter of SETI, who simulated listening to the Earth
from Proxima Centauri, has proposed another challenge for this
strategy of mine: predict at what intervals any pulsed signals are
most likely to be emitted. The periods of the pulsars can serve
as direct standards, enormously simplifying the calculations to be
carried out by NASA's signal detector.

I may, perhaps, have rather let myself go in giving all these figures,
and too many details. This should, however, give you a better idea
of the meticulous care and attention to detail that investigators in
this field need to take in attempting to add a small brick to the vast
edifice of scientific knowledge.

CAN ET LISTEN TO EARTH?

If NASA's system and a large radio telescope were transported to
another planet, somewhere else in our Galaxy, would they be capable of discovering artificial radio emissions coming from Earth?
At a recent SETI forum, John Billingham, head of NASA's SETI
program and now Senior Scientist at the SETI Institute, gave figures for the two methods of searching (using chosen targets and a
sky survey), for the two types of signal that may be expected (continuous and pulsed) and, finally, for the most powerful terrestrial
transmitters (the Arecibo planetary radar and the American ballistic
missile early warning radars).

Terrestrial radars

The potential ranges at which radars can be detected are enormous; the record is obtained when studying a continuous signal from a target star: 4000 light-years. When we recall that the NASA program included just 1000 stars out to a distance of 100 light-years, simple proportion indicates that the volume with a radius of 4000 light-years contains 60 million stars similar to the Sun.

If there had been sufficient time for signals from Arecibo to travel that far, we could be detected from this vast number of stars. In fact, the signals have reached a distance of some 30 light-years only, which means that we could be detected from just 20 stars. Those who fear that we should not reveal our presence may perhaps be reassured; but the radio pulses from Arecibo are, nevertheless, on their way, and nothing can stop them now. Quite inexorably, they will eventually – in merely twice the length of the Christian era – encompass those 60 million stars! By comparison, the signals from the giant military radar installations have a far smaller range of just 20 light-years.

But the question of range does not include all the factors involved in the problem of whether Earth would be detectable. We also need to calculate the probability that our SETI equipment, transported to another planetary system, would be able to discover the Earth, which is an insignificant body lost in the immensity of space. John Billingham has made the calculation, incorporating the amount of time that the Arecibo radar and the military radars are actually transmitting, and the frequency bands that they use. Arecibo, for example, uses its radar for only 200 hours each year. This therefore favors the military radar installations, which are in continuous operation, and which are also more numerous. All in all, their chances are more or less equal. The final probability calculated by Billingham is, however, extremely low. Our SETI installations would have only one chance in a thousand of detecting a planet like the Earth, even if there were one every ten light-years!

Our optimism over the range of SETI systems has to be tempered by pessimism over the likelihood of encountering an artificial signal. A parallel with gambling is perhaps in order: anyone who buys a lottery ticket may win a substantial prize, but their chances of doing so are extremely low. The remedy is to buy lots of tickets. That is the problem: although we have taken the immense technological step needed to detect radio signals propagating over interstellar distances, we still need the right target, the right frequency, and the right moment. But nothing ventured, nothing gained. It is just as well that this calculation's discouraging aspects were not revealed earlier. At least the technology is now under development, and could lead to major additional advances, far beyond a billion simultaneous channels. This will allow us to buy far more lottery tickets.

There are, moreover, other favorable factors. The calculations assume, for example, that extraterrestrial technologies are at our level. On a purely statistical basis, this is not very likely, and if we detect signals they will come from far more advanced civilizations. Their transmitters might therefore be far more powerful than our own radar installations. When all is said and done, this is something for observation to decide. In physics, it is meaningful to attempt an important experiment which might have only one chance in a thousand of success. There are, however, innumerable cases where pioneers have tackled relatively hopeless experiments, only to obtain outstanding results, thanks to two favorable factors: a factor of 1000 gained from advances in technology; and another factor of 1000 gratuitously added by nature itself, which has frequently proved capable of springing far more astonishing surprises than we expect.

Our television signals

The Earth's radar installations are not the sole sources of radio emission. As early as 1978, Woodruff Sullivan showed that television transmitters could also indicate our presence at a considerable distance. The radio signature would be extremely interesting, not because of the content of the programs themselves, but because of the physical properties of the carrier waves. A system like Arecibo would be able to detect them at a distance of 30 light-years.

Because such transmitters are designed to be picked up by individual antennas, their beams are directed horizontally, and a large fraction of their output escapes into space, and is swept across it by the Earth's rotation. By collecting information about the 2200 television transmitters in service at the time, Sullivan reconstructed the nature of the waves that would be picked up elsewhere. There were three main emission centers around the world: Europe, the United States, and Japan. In 24 hours of listening, an extraterrestrial would pick up six bursts of emission, one each time one of the three zones appeared to rise or set, as seen from their position in space. Naturally, Sullivan would have liked to have had a few million dollars at his disposal to pay for a space probe designed to listen to the Earth's signals, and verify his predictions. He found it easier, however, to use the Arecibo equipment for listening, using the Moon as a reflector. Terrestrial emissions are partially reflected by the lunar surface. When, as seen from the Moon, the USSR was setting, and Europe was about to follow suit, Sullivan observed emission peaks at wavelengths corresponding to channel 8 (191 MHz), with the Russian signals fading progressively over a period of an hour, while the Western European ones increased.

What could extraterrestrials, even with our own modest level of intelligence, extract from the physical study of these bursts of emission? By accurately measuring the frequencies of the peaks received, the ETs would find Doppler effects with velocities of up to 460 m/second, the speed caused by the rotation of the Earth. The events would recur every 24 hours. If they found a motion of 300 m/second (say), taken over 24 hours this would correspond to a circumference of 25 920 km. This would enable the ETs to estimate the size of our planet, which would have to be at least 8250 km in diameter. (In fact, the diameter is larger, 12 756 km at the equator.) Then over the course of a year, they would observe another Doppler effect, corresponding to the much greater velocity of 30 km/second, our velocity in orbit around the Sun. From this they could deduce our distance from our star, 150 million km. Then, from the luminosity and temperature of our star, they could calculate that a globe probably 10 000 km across, orbiting at a distance of 150 million km, would have a temperature falling within the range 0–100 °C. And from that it would be possible to say immediately: liquid water, therefore a macromolecular biology, and thus intelligent life!

That is until, by pushing their detector technology to learn more about us, they manage to decode the television programs that we transmit, and find out a lot more about us ... Find out, in fact, that we destroy ourselves with wars, genocide, nuclear pollution, chemical poisoning, and through destruction of our resources, of stratospheric ozone, and of atmospheric oxygen. At the end of the day, our chaotic and irresponsible management of the planet might compel them to strike us off their list of intelligent beings. Without going quite that far, Sullivan points out that certain seasonal factors may be deduced from our carrier waves. In winter, for example, there are fewer leaves on the trees and radio waves are less attenuated. Sociological, and even political indications may also appear: in certain regions the duration of transmissions are less, or may undergo a sudden decline in volume. During the course of his program of listening to signals reflected by the Moon, Sullivan even detected an American Navy military surveillance radar, sited in Texas, that emitted a megawatt of power with a bandwidth of one-tenth of a hertz.

According to some critics, the phase during which a civilization would use radio waves would be of short duration. (Already, we are using fiber optics more and more for telecommunications purposes.) It would therefore be pointless to bank on their radio emissions to detect any such civilizations. As often in this field, this is pure speculation, which can only be answered by counter-speculation. Have people not envisaged space stations designed to capture solar energy, and convert it into narrow beams of microwaves transmitted to stations on the ground? With power levels of 10 gigawatts and a beam efficiency of 99.9 %, such a power station would radiate significant losses into space, and these would be detectable 100 times as far away as any radar. In any case, once again, because the means are now within our grasp, observation will decide.

11

Habitable zones in the universe

NASA's main official program was intended to monitor the 1000 closest stars resembling the Sun. It would have occupied the first few years of the impressive system's operation, and begun SETI on a major scale. Inaugurated in October 1992, after all the various components had been fully integrated, the search program itself would only have begun after about a year for the essential initial commissioning, and would have finished about the year 2000. This gives some idea of the scale of the enterprise. The idea, which had been around for about a decade, is basically simple. There are, however, three major difficulties: 1000 stars is a tiny number when compared with the billions that ought to be examined; the closest stars are not, by their very nature, the most interesting ones; and finally, basing the search on the stars resemblance to the Sun can only be justified because the Solar System contains the sole example of known intelligent life. The other research program, the whole-sky survey, would have compensated for these weaknesses from the outset. This also began in October 1992, but with a prototype having 'only' two million channels. It would have reached its full size only in 1996, when it would have had 16 million channels, and was due to continue into the next century. As we have seen, NASA's sky survey was finally terminated by Congress in 1994.

Stars with planets

Since the targeted program was devised, our knowledge has increased. Efforts to discover planets around other stars have converged on several promising candidates. Other examples have been discovered by accident, such as the very large planet (or brown dwarf?) that has 12 times the mass of Jupiter, and takes 84 days to orbit the star HD 114762. Imagine Mercury replaced by a Jupiter-like planet two or three times the size of our own, and you will have a good idea of this planetary system. If there is an Earth as well, when HD 114762 sets, the super-planet would be an Evening Star hundreds of times as bright as Venus; it would even be visible to the naked eye as a disk, like the Moon.

The discovery of disks of gas and dust around some 20 stars may indicate the presence of planets; some of these stars are already several billion years old, which would have given evolved life forms time to develop. Such stars are therefore prime candidates for SETI.

There was considerable surprise, in mid-1991, when British scientists at Jodrell Bank announced the existence of a planet orbiting a pulsar. The news was soon discredited by the authors themselves, after discovering an error in their calculations. This must have been a dreadful moment for the researchers, and illustrates yet another reason for not rushing into print and for allowing time for calm consideration. There was yet another surprise when, at the end of 1991, it was announced that American workers at Arecibo had discovered two planets orbiting another pulsar. The possibility of discovering planetary masses in orbit around pulsars is extremely promising and of recent date. It requires confirmation by the location of other possible examples, and also by a full understanding of the underlying astrophysics.

Other methods of selecting targets have been suggested, one – which we have already mentioned (p. 164) – is to use the flash from supernovae to synchronize both the transmission of signals by another civilization and their reception by us. If the other civilization's star lies on a certain easily calculated ellipsoid, the signal's travel time will be such that we will see it at the same time as the flash. Another possibility would consist in monitoring stars that could have already been reached by Arecibo's radar beams. They may already be sending us replies ...

Double stars

The majority of stars in our Galaxy are double stars. An important question as far as life is concerned is whether a planet can have a stable, permanent orbit around two stars. These two stars are already orbiting one another (in fact they both orbit their common center of gravity). Investigation shows that, if a planet passes from the gravitational influence of one star to that of the other, it will be subjected to complex accelerations, and will eventually be ejected into interstellar space or else collide with one of the stars – effectively preventing it from having any biological potential. Double stars might not, therefore, be thought to figure on any lists of potential SETI targets. In certain circumstances, however, a stable orbit seems possible. For example, if the planet orbits very close to one of the stars, the effects of the other, more distant star are weak, and the orbit appears to be stable. Similarly, if the two stars orbit very close to one another and the planet has a distant, circular orbit, the two stars will appear to it like a single body, and the orbit will be stable. So double stars do have to be included in the list of SETI objects.

The case of Alpha Centauri

To illustrate the problem, Daniel Benest, from the Nice Observatory, described at the Val Cenis symposium his detailed calculations of the behavior of a planet in a double-star system. He chose an example that is dear to interstellar explorers, Alpha Centauri, the closest stellar system, at a distance of 4.3 light-years. As Flammarion picturesquely put it: 'An express train that started from here would only arrive at this neighboring sun after an uninterrupted journey lasting nearly sixty million years.' In fact, the closest star is actually Proxima Centauri, which is about 2 % closer, and invisible to the naked eye.

Alpha Centauri is a brilliant star in the southern hemisphere. I am always moved when I look at it and realize that it is the closest stepping-stone to the stars. Seen from Alpha Centauri, our own Sun would also be a brilliant sight. With such a small change in position relative to the distances of the majority of stars, our familiar constellations would appear the same, except that, just to the left of the 'W' of Cassiopeia, an interloper would add to the brilliance of that constellation: that is our own star!

The main component of the binary, Alpha Centauri A, is a true twin of our Sun, whereas B is slightly less massive and cooler. Relative to their common center of gravity, their orbits, which they complete in 80 years, are slightly elongated, like an egg, which is the reason why the distance between them varies between 12 and 36 times the Earth–Sun distance (1 AU). Benest's calculations have shown that, apart from very distant orbits (which are of little interest as far as life is concerned, because they are too far away from the source of heat), stable orbits could exist around A or B, with radii of as much as 3 or 4 AU.

Orbits similar to those of Mercury, Venus, Earth, Mars, and even the asteroids are all stable. In the case of a planet resembling the Earth, as we completed our annual orbit of our Sun's twin, A, we would see B in the distance making its tour of the sky every 80 years, at distances varying between those of Saturn and Pluto. Every six months A and B would lie in the same area of the sky, slightly increasing the brightness of daylight. Six months later, B would light up our nights like 1000 Full Moons. Life could, therefore, appear in the region containing what Benest calls 'habitable orbits.' At present, however, there are no reasons for believing that planets can actually condense in this region, because of the complex tidal effects produced by the two stars.

The role of the atmosphere

Our two neighboring planets, Venus and Mars, orbit at 108 and 228 million km from the Sun, as against our 150 million km. Given that the Earth's temperature is ideal, should we be surprised that it is hot on Venus, and cold on Mars? The closer a world is to the surface of the Sun, which is at a temperature of 6000 °C, the more energy it receives, and the higher its temperature rises until it reaches a certain equilibrium temperature, dictated by the losses that it radiates away to space. When we calculate the equilibrium temperatures, we are far from obtaining the 450 °C that prevails on Venus. This is because of the important role played by atmospheres, which may, at first sight, appear to act as protective blankets.

The situation, however, is very complex. For one thing, a planetary cloud cover reflects the infrared radiation emitted by the planet back down to the ground, and largely prevents the planet from cooling. In addition, because it is white, it reflects a substantial fraction of the solar radiation back into space, thus decreasing the equilibrium temperature. A planetary atmosphere is therefore a double-edged sword: it not only reduces losses by its greenhouse effect, but also acts as a protective shield. Full evaluation of these contradictory effects is very difficult, even in the case of the Earth, where we have plenty of data and powerful computer simulation programs. In addition, the equations are such that their solutions are very sensitive to the initial conditions: they are, to use the term that has recently become widespread, 'chaotic.' To quote the usual example, known as the Butterfly Effect: 'If a butterfly's flaps its wings in Brazil today, could it spawn a tornado over Texas next month?'

We are only too aware of problems such as those raised by agricultural practices, as well as the prediction of atmospheric disasters including the greenhouse effect and the ozone holes. The International Space Year, in 1992, had as its main theme the observation of the Earth by a whole range of coordinated, powerful instruments in space. It should help us to suggest measures to safeguard the only planet that we have. In this context, better knowledge of our sister planets, particularly Venus and Mars, would be equally precious. Study of our neighbors would allow us to make basic comparisons between the planets, and realistic extrapolations of future conditions on Earth. Because of their distance, however, these planets are far more difficult to study. Similarly, not many scientists are involved, and both the observational and theoretical methods at their disposal are significantly inferior. Research is still in the early stages: for example, numerical models are not based on three-dimensional rotating spheres, but are reduced to considering atmospheres above an infinite, flat, static surface! In addition, these calculations need to be able to show how the atmospheres of the planets may have evolved over several billion years, not just over a few thousand years.

Habitable zones around stars

Despite the considerable range of issues that need to be investigated, astronomers involved in SETI would like to carry them out, so that they might more accurately define which target stars merit specific, priority attention. The first investigations were made by Michael Hart in 1978, and were designed to discover the range of distances from the Sun at which temperatures on Earth would allow liquid water to exist for four billion years – a basic condition for the appearance of our form of life. The results were surprising: if the Earth had been just 4% closer to the Sun, or 1% farther away, we would not be here, nor on any other planet in the Solar System. This tiny range of just 5% forms what Hart terms the habitable zone (or ecosphere) around our star. These results decrease the likelihood of finding terrestrial-type life elsewhere in the universe.

The carbon dioxide cycle

Because of the importance of the outcome, the calculations have been repeated with more sophisticated models for the primitive atmosphere, in which the initial presence of carbon dioxide, CO_2, and water vapor, H_2O is more accurately modeled. Because these two gases play a significant role in the greenhouse effect, it is essential to calculate how they have evolved over geologic time. Carbon dioxide is involved in an extremely important cycle: rain charged with CO_2 attacks calcium and magnesium silicates in rocks, turning them into carbonates, which are then deposited as thick layers of sediments at the bottom of the oceans. Plate tectonics carries these sediments down toward the mantle, where they melt, and the CO_2 is eventually returned to the atmosphere via volcanoes. In fact, it turns out that our atmosphere has very little CO_2, and that the amount locked up in rocks would create a pressure of 6 million Pa (about 60 bars) if released. In addition, the CO_2 cycle has an amazing, stabilizing role on terrestrial climate. If the Sun's radiation increases, the Earth's temperature increases, more water is evaporated and more carbonates are formed, thus reducing the atmospheric CO_2, and its greenhouse effect. And vice versa.

The new calculations show that the inner radius of the habitable zone decreases slightly, by up to 5% of the radius of the Earth's orbit. With 100% complete cloud cover, the inner radius might decrease as far as the orbit of Venus. As far as the outer radius is concerned, it is determined by the condensation of CO_2 into carbon-dioxide ice in the outer, colder, regions of the Solar System. The greenhouse effect no longer plays a part, and the limit is just inside the orbit of Mars. As we can see, this makes the situation far more encouraging for SETI researchers!

The Gaia hypothesis

In 1979, James Lovelock advanced the theory that life acts to manipulate the terrestrial climate to its own advantage. This daring theory, known as the Gaia hypothesis, has not been well received, but a certain amount of confirmation has been found, together with some explanation. At the Val Cenis symposium, David Schwartzman, from the Department of Geology and Geography at Harvard University, described a model and various estimates that were based on the assumption that the biosphere played a regulatory role. According to him, microbial activity can accelerate the carbon-dioxide cycle, essentially by acting on the fine structure of the soil, producing a very high surface area per unit volume, in effect by very fine-scale tilling of the soil. Up to two billion years ago, the Earth was still very hot, close to 100 °C. It would have been inhabited by colonies of thermophilic bacteria and nothing else. The regulation mechanisms considered previously do not allow the temperature to drop to 50 °C, the temperature required for eukaryotes and eventually human life to appear. By breaking up the soil, however, the primitive bacteria may have caused the carbon dioxide cycle to become a hundred, or even several hundred times faster, leading to a decrease in the carbon dioxide content, and also in the temperature, thus opening the way for more advanced biological evolution. This model enables us to extend the outer radius of the habitable zone to about 4 AU, nearly out to Jupiter. There are, of course, still some obscure points in this scheme of things that need to be resolved, but it does raise interesting possibilities.

Martian-type planets

The role of the atmosphere in determining the habitable zone has also been studied for a planet that, like Mars, is smaller than the Earth. Essential differences arise, not really because the size is smaller, but because of the fact that plate tectonics does not operate. Detailed examination of the tens of thousands of photographs of the martian surface taken by the Viking Orbiters shows that there are no plate motions on our neighboring planet. Mars is a 'single-plate' planet. Without any recycling mechanism, the carbon dioxide in the primitive atmosphere disappeared very rapidly. Geomorphology indicates, however, that large quantities of liquid water existed near the surface in the first one to two billion years of its existence. In addition, shortly after the Solar System was formed, the Sun's radiation amounted to about 70 % of its current level, and Mars (and Earth for that matter) could not have had liquid water. This is known as the weak primordial Sun paradox. The solution is to ensure that the primitive atmosphere had sufficient CO_2 to produce a significant greenhouse effect. A pressure of at least 500 000 Pa (5 bars) is required, in contrast to the current pressure of 600 Pa (6 millibars) on Mars. Starting with these premises, we find that the CO_2 on Mars disappeared in less than a billion years, and when the temperature at the surface dropped below 0 °C on its way to the equilibrium temperature of −50 °C, the water froze.

If liquid water originally existed on Mars, the question becomes one of knowing whether this state persisted long enough for life to appear. The example of the cyanobacteria forming terrestrial stromatolites suggests that it is possible. But it is also important to recognize that the existence of drainage patterns on the surface of the planet does not necessarily imply that they were caused by rains during the phases when the climate was more favorable than at present. The large areas of chaotic terrain, where catastrophic collapse has occurred, are not conducive to the existence of long periods of conditions favorable to life.

If life appeared on Mars, might it still be able to find ecological niches that contain liquid water? The frozen lakes in the Dry Valleys of Victoria Land in Antarctica are ecological niches that shelter significant biological activity, and offer an interesting model. Scientists at NASA Ames Research Center have made preliminary calculations and applied the Antarctic model to Mars. The thickness of the layer of ice covering the lakes is governed by how fast new ice forms at the bottom and how much is lost at the top. This depends on a balance between the amount of insolation (or sunlight received), the conductivity of the ice, the inflow from summer streams, and the latent heat of freezing. Using their models to estimate the number of days that the average temperature rises above zero, they found that, on Mars, similar lakes could have lasted for 700 million years after the average temperature fell below 0 °C. Once again, this is a preliminary result, but it is still encouraging! These calculations, which were undertaken to elucidate the question of primitive life on Mars, also extend the range of possible planets that could be the abode of life, and show that even a small planet, without plate tectonics, might be a potential site in another stellar system.

Other suns

NASA's short-lived program, and the majority of SETI programs undertaken so far, have examined stars that 'closely resemble the Sun.' (As we have seen, thanks to detailed study of their orbits, certain double stars are also potential candidates.) The vast majority of stars may be classified in a regular succession according to their mass, which completely governs the rate at which their nuclear reactions proceed, as well as their temperature, diameter, luminosity, and lifetime. In practical terms, the mass determines all the characteristics as well as the evolution of a star, and these prove to be extremely sensitive to small changes in mass. Although the mass of most stars falls within the relatively small range of one-fifth of a solar mass to 20 solar masses, the lifetimes range from 100 billion years to tens of millions of years, and the luminosities from about one-thousandth to a million times that of the Sun. The surface temperatures vary less, ranging from 2000 K to about 10 000 K, but enough to produce a range of colors from fairly deep red to light blue.

This range of properties is defined by the star's spectral class, traditionally designated by the sequence of letters OBAFGKM. The Sun is a yellow, G star, with a mass of 10^{33} g, a surface temperature of 6000 K, a diameter of 1 400 000 km, and a lifetime of ten billion years. F stars last for about a billion years, but, although they have large ecospheres because their powerful radiation covers a broad region of space, they are not very interesting for SETI, because advanced life forms would not have much time to develop. In any case, they are less numerous than G stars. The most plentiful stars are those of class M. Their ecospheres are very small, however, because their radiation is weak. What is more, any planets that did occur within the zone would always turn the same side to the star, as the Moon does with us, because of tidal coupling. One hemisphere would therefore be hot and the other frozen, which would cause the atmosphere to freeze on the dark side. This is why, until recently, target stars were only of spectral classes F, G, and K.

Stellar ecospheres

New atmospheric models applied to tidally bound planets of M-type stars, however, appear to suggest that a general circulation may set in between the two hemispheres, reducing the temperature differences and allowing liquid water to exist on the surface, whether the planet is of terrestrial or martian type. If this is confirmed by more detailed studies, it will double the number of targets for SETI. At Val Cenis quite detailed results were presented about the ecospheres of terrestrial-type planets, with stellar masses ranging from 0.5 to 1.25 times that of the Sun. A star of 0.85 solar masses, for example, has an ecosphere extending from 0.5 to 1.0 AU at the beginning of its lifetime, which expands, because of stellar evolution, to 1.1–1.5 AU after ten billion years. A star of 1.25 solar masses has a much larger ecosphere, beginning at 1.4–2.5 AU and extending to 2.4–3.3 AU, at the end of its short lifetime of just four billion years.

We can see what would happen to our own planet if it were orbiting these stars at a distance of 1 AU. Around the first, it would start just at the edge of the habitable zone, but, after nine billion years, because of the heat, we would have to emigrate farther out to martian distances, which would have become suitably warm. With the second star, the Earth would not be habitable initially, but the orbit of Mars would be ideal, although not for long. After a billion years, we would have to emigrate to the region of the asteroids, where, three billion years later, we would witness the final catastrophe that overtook the star. Life in the universe is not simple, which is why it is utterly futile for us, down here, to have been making it uselessly complicated for the last 10 000 years ...

Unexpected habitats

Apart from the habitable zones around stars, where the presence of possible liquid water on planets is determined by their atmospheres and the energy received from the stars themselves, other ecospheres are conceivable, with other sources of heat. Jupiter's second satellite, Europa, a smooth sphere 3000 km in diameter, is probably completely covered in a layer of water 50 km thick, with a frozen surface. It is thought that in the interior the water is kept liquid by a form of gravitational kneading that the satellite undergoes, caused by Jupiter and an orbital resonance with Io, the closest satellite. The tides that are produced would release enough energy to keep the water at depth above a temperature of 0 °C. (On Io, such tidal effects are so strong that they produce active volcanoes erupting an infernal mixture of sulfur compounds.) It is also possible that there still may be pockets of liquid water inside giant cometary nuclei.

The spiral arms of galaxies

I would now, however, like to take you far outside our Solar System, and even beyond the region of neighboring stars. This time, we need to consider our Galaxy as a whole. The story, only recently unraveled, is surprising. The Galaxy consists of a central bulge, a gigantic swarm of stars, strongly concentrated toward the galactic nucleus, around which orbits a flattened disk of stars, gas and interstellar dust. The stars move in circular orbits around the nucleus, like the planets around the Sun, and in accordance with Kepler's laws, whereas the gas and dust are governed by the laws of hydrodynamics, with effects caused by viscosity, wave propagation, and shock waves.

It is this hydrodynamic behavior that is responsible for the existence of density waves, i.e., for regions where the density of the gas is higher. In addition, these density waves are located where they satisfy the hydrodynamic equations, which in fact force them to occur in two regions that spiral out from the center. Finally, when the gas approaches such a spiral region, it is accelerated by the higher density material and creates a shock-wave, which initiates the formation of new stars within the gas. It is these stars, some of which are extremely bright, that illuminate the spiral regions, creating the magnificent arms that cause spiral galaxies to appear like exotic jewels set in extragalactic space.

It is a nice picture, but even better is yet to come, and this takes us back to SETI. The laws of hydrodynamics force the spiral arms of dust and gas to rotate as a whole, without distortion (rather like the wake behind a ship). They take 200 million years to complete one revolution around the galactic center. The stars themselves, however, orbit more slowly, the farther they are from the galactic center, just as the planets do, with periods ranging from 88 days for Mercury to 250 years for Pluto. But here is the extraordinary thing: the Sun, which lies 30 000 light-years from the galactic center, also takes 200 million years (to within a few per cent) to complete its journey around the galaxy.

Habitable regions in the Galaxy

This extraordinary coincidence attracted the attention of L. S. Marochnik and L. M. Mukhin, from the Institute of Space Studies in Moscow, and B. Balázs, from Eötvos University in Budapest. They described their results at a bioastronomy symposium held at Balaton in Hungary, in 1987. First of all, this situation means that a star like the Sun rarely crosses a spiral arm. Currently, it is rather more than halfway between the Sagittarius Arm and the Perseus Arm. It left the former 4.6 billion years ago and will encounter the latter in 3.3 billion years. So the Solar System was born within the region of active star formation in the Sagittarius Arm. Expanding on the ideas of Iosif Shklovskii, the authors believe that when our Solar System next crosses the active region of the Perseus Arm, it will pass close to a number of supernovae. If we pass within 30 light-years of one, the cosmic ray flux will increase 100-fold, giving rise to such radiation doses that the whole human race will disappear, unless its rate of increase can compensate for the number of deaths. The authors estimate that in the Paleolithic the population doubled in 200 000 years, whereas now such doubling takes just 30 years, which would compensate for the future effects of supernovae. It will be essential for our rate of increase to decrease in the near future, however, because life would become extremely difficult if the overall human population exceeded ten billion. Consequently, in 3.3 billion years, our civilization will be destroyed.

Galactic zones and advanced life forms

These results suggest that only stars orbiting at about the same speed as the spiral arms can acquire civilizations comparable to our own. The calculations show that to escape the devastating effects that arise when crossing a spiral arm for a sufficiently long time for such civilizations to arise, the parent stars need to lie within an astonishingly narrow, galactic habitable zone. This is a ring, some 1500 light-years wide, lying 30 000 light-years from the galactic center. This 'advanced-civilization corridor' contains about a billion stars, only 100 million of which might have habitable (stellar) ecospheres. Suitable SETI targets, therefore, are to be found only in this 'ring of life.' B. Balázs has studied the distribution of these stars on the sky; it follows the Milky Way, reaching a maximum tangent to the Sun's orbit around the center of the Galaxy, decreases in the direction opposite to the galactic center, and is zero directly toward it. In the two most favorable directions, one square degree should contain 1000 candidates, as against 1 or 0 in the other two cases.

These galactic factors, taken together with those I have proposed in favor of using pulsars, are therefore essential in trying to develop suitable SETI strategies. In addition, because they are applicable to all spiral galaxies, they offer a new perspective on the complex, astrophysical conditions that are involved in the appearance and evolution of life in the universe. Nearly a decade ago, when I gave up work on abnormal galaxies to devote my time to SETI, I thought about writing a paper on 'The impossibility of the appearance of life in clumpy galaxies,' because these are absolutely swarming with innumerable, fatal supernovae ...

12

The day we make contact

Science is normally well received by the general public, but there are still some ferocious opponents. In principle, if A = B and B = C, then A = C. Everyone agrees about such a 'result.' But science is constantly evolving. It is always at the frontiers of knowledge, and is always trying to push into unexplored territory, which seem to be an impenetrable thicket. The explorers, like everyone else, are prone to prejudice. Frequently, too, when they come across a sensational find, they are unable to resist the temptation to appropriate it to themselves, and usually much too quickly. That is where the error lies.

It is therefore of the utmost importance, given what is at stake in SETI, that scientists working in this field constantly examine their own motives and exercise restraint. We only have to recall the matter of CTA 102 (p. 150). Nicolai Kardashev's superior offered *Pravda* a scoop, and its effects were such that Charles de Gaulle, after meeting Jean-François Denisse for the first time a month later at the Nançay radio telescope and discussing the matter with him, immediately took the opportunity to send a diplomatic warning to the Soviet administration that was intended to dampen their enthusiasm. We should also remember the signal that Frank Drake picked up from an airplane, when he carried out his pioneer observations. A similar vigilance needs to be exercised over dissemination of the information. We need to take care that we do not get involved in pointless arguments or pseudo-scientific debates. It is precisely these considerations that have prompted people to draw up a protocol that lays down the conditions that should be followed in making any public announcement about any possible 'contact.'

An extraterrestrial signal belongs to all humanity

Another motive, fully as important, also played a part: we need to avoid any tendency for such a detection to be appropriated by individuals or institutions. The scientists engaged in this research, in particular the radio astronomers, believe that, if a signal is received, it should be considered as part of the cultural property of all mankind, from the very moment it is detected. In addition, contact, because of its potential significance, involves the whole of humanity. Finally, it seemed to us that drawing up a protocol governing the announcement would be a vital element in rallying public opinion and demonstrating the serious nature and the potential interest in SETI.

It all began with the SETI Committee of the IAA, which invited the submission of ideas bearing on the question from a wide range of disciplines, including science fiction, sociology, law, and politics. These contributions were published in a volume of *Acta Astronautica*. They discussed the reception and verification of a signal; the announcement of the discovery and its impact, and finally the legal aspects of this contact. At the very end came the question of who should speak for Earth. After several preliminary versions, a text was adopted by the IAA and the International Institute of Space Law in 1989, by the Bioastronomy Commission of the International Astronomical Union (IAU) in 1991, and by the Committee on Space Research (COSPAR) and the International Council of Scientific Unions (ICSU). This declaration has a number of points in common with the one that governs the protection of the planets, which was introduced by COSPAR in 1984. The final form of our text was also inspired by the Treaty on the Exploration and Usage of Space, which requires member states to inform the Secretary General of the United Nations, as well as the public and the international scientific community, about the nature, the conduct, the places, and the results of their activities in the exploration of space.

Here is the preamble of the SETI Declaration:

> We, the institutions and individuals participating in the search for extraterrestrial intelligence,
>
> Recognizing that the search for extraterrestrial intelligence is

an integral part of space exploration and is being undertaken for
peaceful purposes and for the common interest of all mankind,

Inspired by the profound significance for mankind of detecting
evidence of extraterrestrial intelligence, even though the probability of detection may be low, ...

Recognizing that any initial detection may be incomplete or
ambiguous and thus require careful examination as well as confirmation, and that it is essential to maintain the highest standards
of scientific responsibility and credibility,

Agree to ...

There follow several points about verification by the discoverer,
then by the signatory parties, with the establishment of a network
for continuous monitoring of the candidate signal, before any public
announcement is made, except to the discoverer's national authorities. This would be followed, in the case of a credible signal, by
announcement to other observers, to the Secretary General of the
United Nations, and to various international scientific unions; and
finally, if the case is confirmed, dissemination of the information
by every scientific channel and through the media. Also discussed
are the monitoring, recording and dissemination of the signals received, and their protection from radio interference, which is to
continue under the SETI Committee of the IAA, in coordination
with the Bioastronomy Commission of the IAU. In addition an international, multidisciplinary committee will be created to serve as
a focal point for the longer-term analysis and public dissemination
of information. Finally, it is laid down that the discoverer should
have the privilege of making the first public announcement. No
a priori response will be sent, and the IAA will be the depository
of the Declaration. This outline was given at the Val Cenis symposium by the three principal instigators of the Declaration, John
Billingham, Michael Michaud (Scientific and Technical Advisor at
the American Embassy in Paris, later in Tokyo), and Jill Tarter, in
cooperation with colleagues in various countries: Argentina, Austria, Czechoslovakia, France, Italy, the Netherlands, Poland, and
the United States.

The scientific and media points of view

This initiative may be thought to be premature, given the low probability of detecting any signals. After all, would there not be enough time to react when anything happened? Some 30 top American science journalists were asked this question by Andrew Fraknoi, the Executive Director of the Astronomical Society of the Pacific. All of them felt that contact with an extraterrestrial civilization would be the most important scientific event of our time. One-third thought that they would hear the news through a press conference or by rumor, and considered that it would be difficult to check. Their confidence would depend on the reputation of the person making the announcement. They would liken it to the discovery of pulsars, and would try to contact the person making the announcement and their colleagues, Sagan and Drake being at the head of the list. If they failed to do that, they would have recourse to theologians, political leaders, and the White House.

Under such circumstances, the normal methods used for scientific communication would be unable to function; they would be too slow to react, and would be overtaken by general panic, haste, and sensationalism. This is why the journalists clearly expressed their desire to see the information distributed as widely and as fairly as possible, in accordance with a strategy worked out in advance. This would be the only responsible way of disseminating the information.

The researchers themselves took the same view. In a larger study, Donald Tarter, professor at the Department of Sociology of the University of Alabama at Huntsville, sent 300 questionnaires to various scientific media and to SETI researchers in 20 different countries, with a view to determining their perception of the importance, the level of information, and the credibility of SETI, and to judge the best announcement strategy. His conclusion was that:

> The discovery of ETI would mark the entry into a new age of human understanding of the cosmos. It could lead to a complete reassessment of our place in the universe and the nature of life as a natural evolutionary process. A systematic search may take decades or centuries before any evidence of ETI is found. Indeed, there is no assurance that the object of the search exists, therefore, the search may never be successful. On the other hand, we must

be prepared for the possibility that the discovery of ETI could be made at an early stage in the search.

To both researchers and the media, SETI is one of the most important projects in the history of science. Its potential impact surpasses that of a mission to Mars, a lunar base, and an orbiting space station – to take just projects involving space. Nevertheless, this extraordinary adventure is frequently denigrated both by scientists engaged in other fields of research and by political authorities, not to mention the general public, which most of the time has no conception of the serious nature in which it is regarded, nor of the rigorous methods, the scrupulous, meticulous care, and the critical evaluation that are involved in the whole affair.

With regard to the historical fame that the first discoverer will acquire, the researchers felt that common sense and critical analysis would prevail over haste. In any case, it will, in practice, be the SETI computer that will raise the alarm. After all the necessary initial tests to determine, by a progressive process, that it is a real alert, several scientists at various observatories will be fully involved. A good example is given by the physicists, who use far greater pieces of equipment: the announcement of the discovery of intermediate vector bosons was signed by dozens of names.

Donald Tarter has proposed the establishment of a Verification Committee, and this idea has been overwhelmingly supported. This would help with interpretation and analysis before any announcement is made. In this context, I have suggested that a global SETI network should be set up, which could provide technical support to the committee. Donald Tarter has also analyzed the probable reactions of the general public. Initially, the main impression would be one of immense interest and excitement. Then, it was said, there would be a considerable level of confusion and incredulity. Fear and shock arising from contact would appear last. The survey revealed the possibility that there would be angry reactions to the news and questionable exploitation of it by certain groups such as 'UFO cults, religious fanatics, and even commercial enterprises ... The Committee would help to avoid undesirable, and possibly dangerous, consequences,' Tarter concluded. According to Alain Cirou, Chief Editor of the French astronomical journal *Ciel et Espace*:

For the Voyager missions, JPL set up an exemplary system of communication ... Every day, a press conference brought together all of the representatives of the press and scientists. Everyone accepted that not everything would be said, not everything shown. But the meeting was fixed, credible, and 'official' without being overwhelming. Photographs and a press release were distributed. It was true 'communication in a crisis.'

He concluded: 'A SETI communication will be communication in a crisis, and the appropriate methods should be adopted. Which is why we need to study the matter.'

A natural or an artificial phenomenon?

To give an idea of the exacting nature of science, it may be likened to that required in the administration of justice. Without fairly serious trials, science would never amount to anything; it would simply be an incoherent series of rumors. In this respect, the confusion between SETI and Unidentified Flying Objects (UFOs) is disastrous. *A priori*, astronomers believe that interstellar travel by material objects is possible, and that some might even fly into view. The basic concept of UFOs – or rather of objects from another civilization – cannot be rejected out of hand. It cannot be denied, however, that to this day no seriously documented case exists. The reported sightings of UFOs are just that, sightings, and there is an enormous difference between a mere sighting (regardless of the reliability of the witnesses), and a proper scientific observation.

The 'success' of UFOs is perhaps partly due to the ease with which rumors spread. A psychologist at Fichburg State College, in Massachusetts, considers rumors as being like opportunistic viruses that infect information, and which propagate thanks to the anxiety that they themselves produce, and to the mutations that they undergo, which adapt them to new situations. Certain myths may persist for centuries, simply changing their hosts. Are UFOs just one of these myths, muddying the troubled waters of the subconscious mind that is worried about its place in the universe, its origin, and its destiny? Are they merely one of the horde of visions that have haunted our minds for millennia? Whatever happens, if a single UFO is truly identified, it will become part of the common property of scientists the world over.

A recent discussion at the Val Cenis symposium throws an interesting light on the attitude of scientists to UFOs. The subject was the question of what is meant by artificial or natural in the SETI context. Another aim was to try to analyze our attitude with regard to the evaluation of such signals, particularly with respect to the Russian goal of investigating signals that might come from extraterrestrial astro-engineering works.

At one time, influenced by Shklovskii, one of the first Soviet astrophysicists to promote SETI, the attitude was to consider the hypothesis of an artificial origin only after having refuted all natural hypotheses. The attitude was to assume that phenomena were 'natural.' This would prove one's honesty in not wanting to bias research in favor of extraterrestrial civilizations, which were not generally accepted at the time. In a discussion at Val Cenis, however, the Ukrainian astronomer V. V. Rubtsov accorded equal status to the natural and artificial aspects of the famous explosion that occurred in 1908, and flattened trees over an area tens of kilometers across in the Siberian taiga near the river Tunguska.

For 40 years, the predominant explanation was that the event marked the fall of a giant meteorite. In accordance with the proper scientific method, this could be confirmed by searching for debris buried in the local marshes. In 1946, a Soviet engineer and science fiction writer proposed the theory that some disaster had overcome a nuclear spaceship as it neared the end of its journey, and that therefore the area should be checked for traces of radioactivity. Then in 1958, an expedition by the Soviet Academy of Sciences showed that the explosion took place in the upper atmosphere and that we were not dealing with a normal meteorite. Next, in 1961, there came the theory that it was the explosion of a small comet, caused by atmospheric friction. In 1975, when no traces of radioactivity caused by fission, fusion, or antimatter were found, the theory advocating a nuclear explosion was restricted to an explosion of exotic material. In an attempt to make further progress, the idea was introduced in 1966 that there had been a change of course at the end of the flight. The path was from east to west, agreeing with the analysis of the fallen trees, whereas eyewitness accounts, collected in 1920 in a less rigorous fashion, suggested that the object arrived from the south or south-east. In fact, arrival from the east was reported by other eyewitness accounts from as far away as 1000 km.

What can we conclude from all this? What we can say straight away is that scientists are not, in principle, opposed to the idea of UFOs. As we can see, this case resembles criminal investigations in many respects, and it is well documented. Nevertheless, the jury would be unable to reach a verdict, although they would probably incline toward the 'natural' explanation: that a portion of a comet, amounting to some million tonnes encountered the Earth's atmosphere, and explosively disintegrated at an altitude of 8 km, all from perfectly natural causes.

Here is another, more recent, case. Early on the morning of 26 January 1992, two astronomers at the European Southern Observatory, A. Smette and O. Hainaut were finishing their night's observations at La Silla.

> We left the dome and stopped to have a look at the beautiful dawn in the eastern sky over the Andean mountains. We were trying to find all the planets visible in that part of the sky ... It was Alain who first noticed a bright, diffuse object in the south-east direction. It was moving towards north, about 15° over the horizon. We could follow it during about three minutes, then it was no longer visible as the morning sky became brighter and brighter ... Through binoculars, it had a bright condensation ... surrounded by a 2° wide, circular nebulosity ... [They were able to exclude a satellite, aircraft, meteor, or a cloud of lithium or barium.] Weighing all the facts, we are most inclined to believe that what we observed was actually a natural object, passing very near the Earth, although we would not entirely exclude that it may have had an artificial origin.

A photograph with a 20-second exposure clearly shows a trail. It should be noted that these astronomers were prepared to consider various possibilities, but, knowing their job as observers of the sky, they did not attempt to guess the distance, unlike the majority of UFO addicts, who generally make this fatal mistake. Smette and Hainaut considered many possibilities, from an aircraft flying at a height of a few kilometers, to a comet moving some tens of thousands of kilometers away. Depending on its true distance, which is unknown, the object, whose apparent diameter was all that was known, might have had an actual size ranging from 100 m to 1000 km. An utterly fundamental difference, if one were to try to draw premature conclusions about the nature of the UFO! (In fact this 'UFO' was later identified with waste dumped from the Space Shuttle!)

Be that as it may, the harm done to SETI by this confusion with unidentified flying objects is very considerable. In 1990, for example, the US House of Representatives voted for an amendment reducing the amount of money allocated to NASA's project, following a heated speech by one of the congressmen. According to him, America should not waste precious dollars searching for little green men. He added that cutting of funding would prove that there was still intelligent life on Earth. It could be said, however, that it is not SETI that needs to be financed, but SCI: the Search for Congressional Intelligence! A statement by the SETI Institute did restore the balance and swayed those who were undecided. Luckily, the Senate reinstated the money requested by the President. Unfortunately, in 1993, confused Congressmen again attacked SETI and this time succeeded in denying NASA any funds to continue with its own program or to support any other institution engaged in SETI. It hardly needs to be said that it is vital, both for serious scientific research, and for ensuring that the general public is provided with sensible information, that a clear line should be drawn between SETI and UFOs.

Should we reply and start a dialog?

Proposition 8 of the SETI Declaration stipulates that 'No response to a signal or other evidence of extraterrestrial intelligence should be sent until appropriate international consultations have taken place. The procedures for such consultations will be the subject of a separate agreement, declaration or arrangement.' The first person in official circles to raise the problem of any possible reply was Donald Goldsmith, Director of Interstellar Media, at Berkeley in California. At the symposium in Balaton in 1987, he presented a paper entitled: 'Who will speak for Earth?' That same year, G. C. M.

Reijen, from the Faculty of Law at the University of Utrecht, set out his views on the basic elements of a reply.

To many people who know little about SETI, such preoccupations seemed, and still seem, extremely premature. Nevertheless, if a reply has first been accepted by a portion, albeit small, of society, the greater the chances are that the matter will be taken seriously. To promote a bilateral discussion with extraterrestrials, the first reply needs to be clear, intelligible, and interesting to the other civilization, rather than a nebulous, confused message that is hard to understand.

But is it really possible to consider starting a dialog, if a message comes from a body 100 light-years away, when it will require 200 years for a round trip? Why not? On the galactic scale, we need to take a broad view. Are we not still interested in Homer, 2000 years later? Is not *Homo erectus* an example of a culture that lasted a million years, and, without going as far back, did not the Magdalenians, those highly talented artists, hold sway for 10 000 years? If we look beyond the short span of a single human being's lifetime, it is not unreasonable to envisage interstellar dialogs that outlast it by a wide margin.

During the first few years of its intended operation, NASA's SETI program might have captured signals originating at what are, in galactic terms, relatively modest distances, but which correspond to travel times ranging from less than 100 years to hundreds of centuries (in the sky-survey mode), and which may thus be feasible for individuals or civilizations that last as long as the Magdalenians. If we want to start a dialog, we shall, according to Goldsmith, be in competition with a lot of other 'correspondents' that have already contacted the civilization that is transmitting. A poorly prepared reply will not have much chance of initiating sustained communication. Our initial response may play a decisive part in starting a dialog; we need to have the most intelligent stock of knowledge to put forward, which could not be done in confusion.

From the technical point of view, once we have received a signal, we will know the frequency, the bandwidth, the direction, and probably the distance (from astrophysical methods) and thus the transmitter's power. Quite possibly we will also know the type of coding used. This information will be used to the full by the technicians when it comes to sending a reply. The messages already sent by the Arecibo planetary radar have ensured that we have the necessary technology.

It was not until 1991, at the General Assembly in Buenos Aires, that the IAU was able to adopt the SETI Declaration. In the meantime, the SETI Committee of the IAA, true to its pioneer spirit, took the initiative at the Forum held in Dresden in 1990. Michael Michaud, John Billingham, and Jill Tarter presented the first draft of a White Paper entitled 'International Policy on a Reply from Earth'. This states that a reply should be sent in the name of all humanity; the decision to reply or not should be taken by an appropriate international body; and the content of the response should reflect an international consensus. The potential significance of the reply is such that 'Many of the issues are not primarily scientific in nature: they are social, philosophical, and political. They involve space law and international law. They are therefore more suited to the United Nations than to international scientific or space societies.'

This task will be handled by a subcommittee of the SETI Committee itself that was set up at that meeting (consisting of six Americans, an Argentinian, a Frenchman, a Pole, and a Czechoslovak), and then considered by the International Institute of Space Law, and the IAU Bioastronomy Commission. It will then be passed to the United Nations, via the Committee for the Peaceful Uses of Outer Space.

In the conclusion of their report, Billingham offers a broad perspective of the issues involved:

> by holding a conversation with another civilization, succeeding generations of mankind may gain a wealth of new knowledge; this could range from an understanding of the past and the future of the Universe to physical theories of the fundamental particles of which the Universe is made, and to new biologies. We might be able to converse with distant and venerable thinkers on the deepest values of conscious beings and their societies. We may even become linked with a vast galactic network of extraterrestrial civilizations with unimaginably rich cultures.

Without looking as far ahead, more immediate views may, according to Vladimir Kopal, a lawyer from the Outer Space Affairs Division of the United Nations, encourage humanity as it begins its uncertain path into the future:

> Elaboration of the principles and norms to govern relations between the international community of our own planet and other intelligent communities in the universe would add a new dimension to the present body of outer space law. At the same time this new approach might exercise a beneficial influence on relations between nations and peoples of the planet Earth.

Elaborating on these first statements, I looked for a pragmatic approach which could lead to a consensus. Simple requirements form a lead:

> not to try formulating a special text, as proposed by Reijen,
> keep due account of Goldsmith's concern about potential human censors,
> get at the start the wider consensus possible,
> the initial text contains as much as possible valuable information,
> it is completely formulated and intelligible in its coding and transmission.

Then the answer is quite obvious: just send *them* an encyclopaedia! The *Britannica* or the *Universalis* answers all the criteria:

1. their vast diffusion, then their acknowledgement, insure their recognition by a large portion of the Earth society and the best chance for a wide consensus;
2. they are the output of a very wide section of all human learned people, from all disciplines, with all philosophical points of view and nationalities, aiming at insuring to them an impartial, factual and scientific spirit;
3. they contain a maximum amount of valuable information in a single piece of work, this being their *raison d'être*;
4. they are essentially a linear string of typographic signs (the text) and a set of bidimensional arrays of pixels (the illustrations) whose coding is elementary. The alphabetical coding can be deciphered using just a few pages, as well as the grammatical structures. The illustrations

are also obviously decodable by any ETs using bidimensional information from their own environment. The coupling between text and illustrations will easily provide information nearly *ad infinitum*. From them the structure of the access to the corpus will also be easily decodable;

5. they are already available, always in an up to date version, even to a point where themes like 'bioastronomy' are included;

6. they are easily transmissible; the text can be coded into one billion bits while the illustrations, comprising billions of pixels, can be coded with 100 billion bits or so. With a 10 MHz transmitting band, the complete transmission would take only 3 hours.

Epilog:
Terrestrial destinies and cosmic perspectives

How the discovery affects us

It is obvious that the discovery of the New World by Christopher Columbus altered completely the views of the inhabitants of the Old World. This was despite the fact that 2000 years ago, certain Greeks knew that the Earth was round, that it was a sphere 10 000 km in diameter, and that it was floating in space or resting stationary in the center of the universe. Similarly, the Vikings had arrived in America half a millennium before the famous flotilla led by the *Santa Maria*; and educated navigators, well versed in their craft, knew that sailing toward the west one would return from the east. While the facts had not been demonstrated in practice, it remained just an interesting and stimulating conjecture. At the most it was a working hypothesis, just like the one that forms the basis of SETI: other New Worlds do probably exist beyond the cosmic seas. It is an extraordinary prospect. But it will not be until we have received a signal as a result of our observations that we will be convinced, body and soul, by the universal vision of a Single Cosmos, like the universal vision of a Single Earth that Columbus brought about.

Although the discovery in 1492 opened new horizons to Europeans, it also led to the almost complete destruction of the civilizations and peoples of the New World. Will we be able to resist the temptation of allowing ourselves to behave in a similar way again?

Let us hope that the commemoration of the discovery of America may contribute to a new feeling of fraternity and invite respect, in future, for all forms of life.

Our destructive behavior has its origins in the beginning of the Neolithic, some ten thousand years ago. It was not, therefore, caused by physiological factors, because *Homo sapiens* did not take any particular evolutionary step at that period. On the contrary, agriculture and the domestication of animals were discovered, and people began to build up reserves of food. The population increase accelerated, leading to closer proximity with neighbors and the disappearance of the earlier, protective, dispersion of families and tribes. What could be more 'natural' then, for this impetuous being than to attack and kill one's neighbors to obtain their stocks?

Henri de Lumley, a professor at the Muséum d'histoire naturelle [Natural History Museum] in Paris, and Director of the Institut de paléontologie humaine [Institute of Human Paleontology], notes:

> With the process of settlement, agriculture, and the rearing of stock, the accumulation of worldly goods appeared and with it, its corollary: covetousness. The Hypogée de Roaix, in the French department of Vaucluse, with its 'war grave' clearly illustrates this phenomenon ... This extraordinary pile of perfectly articulated skeletons sustains this theory, because the bodies were obviously deposited at the same time. This idea is reinforced by the presence of several arrow-heads still stuck in the bones ... The arrangement of the bodies is haphazard, with men, women, and children all intermingled and piled up without any sort of order, often upside down.

This indiscriminate massacre dates back to 2000 BC, in the middle of the Neolithic. Our appalling behavior is therefore of comparatively recent origin.

Over the past 10 000 years, the progress in human technology has been astonishing. Although, in principle, these advances could have been used in a perfectly realistic and practical way solely to improve our condition, they have, unfortunately, been placed at the disposal of our tendency for pillage and violence. To such an extent in fact, that the whole globe is now threatened with ruin. Is this the natural order of things? Is this terrible behavior that has characterized the human race since the Neolithic, merely a specific instance of the general fate of the universe? The latest, and most gigantic story of destruction in the universe may make us believe that this is so. Here it is.

The Hubble Space Telescope was just the first in a series of major orbiting astronomical observatories that NASA, as principal project manager, planned to launch to study astrophysical processes in the universe. Although the first did not initially prove to be satisfactory, the second, in contrast, has been an immense success. This is the Compton Observatory (named after the great American physicist), which was designed to orbit the Earth and observe gamma-rays from the universe. It was launched in 1991 from the Space Shuttle, and has a mass of 13 tonnes, greater than that of the Hubble Observatory, and carries, among other experiments, one known as BATSE (Burst and Transient Source Experiment), designed to measure gamma-rays, which are photons that are far more energetic than X-rays.

For years, brief bursts of these gamma-rays have been observed. They last just a few seconds, but are extremely powerful. Although the directions from which they arrive are still poorly defined, it was thought that they were slightly more concentrated toward the plane of the Milky Way, and that they might be emitted by very unusual stars within the Galaxy. Some people had even suggested that they might be caused by the impact of comets onto their central star, which had become a neutron star. This would release an enormous amount of gravitational energy instantaneously, and some of this might be found in the form of gamma-rays. BATSE has, however, scanned the whole sky, and has produced some preliminary results. At the European Conference for the International Space Year, held in Munich in April 1992, G. J. Fishman, from NASA's Space Science Laboratory at Huntsville, told us that the sensitive detector had found 300 gamma bursts, averaging approximately one per day. The most surprising thing was that their distribution over the sky appeared to be perfectly isotropic. There was no tendency to follow the Milky Way, and equal numbers were found in every direction. It is no longer possible to argue that the objects responsible for these bursts are located in the body of our Galaxy. They either come from stars that are extremely close to us, and which therefore appear isotropically distributed over the sky, like the stars in the immediate vicinity of the Sun, or else they are found far out in intergalactic space, where they are also isotropically distributed.

There is much at stake here, because if the sources are nearby their intrinsic power is low. If, however, they are at extragalactic distances, they are radiating a vast amount of energy, at a hitherto unknown rate, indicating the existence of new, violent astrophysical processes. This question has now been settled: the directions do not appear to be associated with the well-established positions of nearby galaxies. In addition, by using methods that count the number of events as a function of the energy received, as was done a long time ago with the powerful radio sources discovered when radio astronomy first began, it turns out that the bodies from which the gamma-ray bursts originate are at distances comparable to those of quasars, i.e., billions of light-years.

Their intrinsic energy is then so great, and the interval in which it is released is so short – some bursts last less than one-hundredth of a second – that one of the possible explanations that have been suggested is that the bursts arise when neutron stars fall into black holes! When we think that a neutron star and a black hole are both remnants of a catastrophic end to stellar evolution, and that these two remnants combine to give yet another remnant that is even more extreme, in an even more irreversible catastrophe, we do not get a very optimistic vision of the ultimate fate awaiting our universe. Should we then feel that our own human catastrophes are in the order of things?

Not necessarily. An incredible series of 'miracles' did, in fact, enable humans to emerge and develop intelligence. In getting as far as we have, we have been constantly threatened or, even worse, completely ignored, by the universe. Each time, however, favorable circumstances have arisen when the fate of the universe has taken a different turn, such that, in the end, we are here. This takes us to the very heart of what is known as the anthropic principle, which was originally, and correctly, stated by Brandon Carter, CNRS Research Director of the Paris Observatory at Meudon.

This principle was soon expanded, distorted, and used for teleological ends (i.e., as an argument for the existence of God): people have even maintained that the whole of the vast cosmos has evolved over a tremendous span of time so that the human race could emerge. This is obviously highly presumptuous, even if we fall back on placing the responsibility on some infinitely powerful imaginary being. In addition, there is nothing to support this controversial interpretation. The real anthropic principle is far more modest: if the universe seems to have evolved so that we may exist, it is solely because we do exist. If it had not done so, we would not be here, and we would not be able to state that the universe could have evolved in some other fashion, particularly in an unfavorable way. The principle is nothing more than an observational fact, or even a selection effect, pure and simple, that puts things in their place. It throws some light on a form of internal inevitability in the universe, but one that is established after the event, and that is certainly not deterministic (i.e., it is not one that is fully determined by prior events and is thus inevitable).

In fact, there is nothing to stop us imagining a whole range of possible evolutionary paths that the universe might follow. Our universe, the one that has been such that we are able to exist, is only one case among many others. This is why, throughout this book, I have frequently used the expression 'our universe' rather than 'the universe.' The remarkable thing about recent developments in the theory of the Big Bang, specifically the 'Chaotic Big Bang', is that they leave open the possibility of an indefinite number of different universes, with different properties, and different fates. It is easy to imagine that the vast majority of these universes are unsuitable for the emergence of intelligences capable of marveling at their universe, of thinking that it is the only one, and of deducing, quite incorrectly, that their particular universe evolved so that their intelligence should appear. In this view, the true universe consists of this whole indefinite number of different universes.

Some people, in taking the anthropic principle to an extreme, believe that they see cause for hope that the universe works on our behalf: the human catastrophes of the past 10 000 years are not in the true order of things. To my mind, it is better not to deceive ourselves. In any case, long-term optimism is not forbidden, according to Freeman Dyson, whose fertile mind has been working at the famous Institute of Advanced Studies at Princeton for many years. He is the author of the concept of the longest period (the 'Dyson Age') ever considered in physics: 10 to the power of 10 to the power of 76 years ($10^{10^{76}}$)! This is such a large number of years that to write it down we would require the numeral 1 followed by as many zeros as there are protons in all the billions of observable galaxies. If, at this extremely far distant time in the future, our universe still exists, black holes will have long since evaporated into pure radiation in the Hawking process, which, for its part, requires 'only' 10^{100} years.

By the Dyson Age, the universe will be nothing more than expanding space, cold, sparsely populated by a few rare photons, neutrinos, and some other, even rarer, particles. The laws of physics as currently understood still allow the possibility of activity of unlimited richness: there will always be something new to discover. 'I have found a universe of unlimited richness and complexity,' wrote Dyson, 'a universe in which life can continue indefinitely, and can reach neighbors across unimaginable vistas of space and time ... There are valid scientific reasons for seriously considering the possibility that life and intelligence could succeed in modifying the universe for its own ends.' Such views, although they remain within the domain of the possible, are only speculations. They do, however, have the merit of allowing us to think that not all is lost, and to nurture the hope of being able to counter our destructive behavior.

Bioastronomy

The young field of bioastronomy is only a dozen years old. It was in Patras, in 1982, under a brilliantly blue, Greek summer sky and by a deep blue sea, that the IAU's Commission 51 was born, with Michael D. Papagiannis as its President. Since then, bioastronomy has identified and opened up a vast expanse of territory, one in which it is easy to lose one's bearings, but where various claims have been carved out, specific stages and milestones have been established, and reasonable theories have been devised to map out the territory.

What discoveries await us? Will we succeed in building an organic molecule that is capable of reproducing itself? Will we discover fossil life on Mars? Will we actually manage to capture an artificial signal? These three questions recall the three basic factors that have fostered the birth of bioastronomy: macromolecular biology, the exploration of space, and radio astronomy. There are many other points that keep us in suspense: will we find a proper, terrestrial-type planet orbiting another star, adenine on Titan, or amino acids in interstellar space? (As regards the last point, the answer looks like yes! To a 90% confidence level, the smallest amino acid, glycine, may have been detected by radio astronomers in the Sagittarius B2 star-forming region near the center of the Galaxy.) Will we be able to build receivers capable of monitoring billions of channels simultaneously? Will we be able to return cometary soil samples to our terrestrial laboratories? Will we be able to manage the announcement of the discovery of any eventual signal successfully?

In the longer term, taking an even deeper and more fundamental view of things, will we succeed in unraveling the various twists and turns that have occurred during the evolution of the universe, life, and intelligence that have brought them to their current state, after some 15 billion years? Will we be capable of finding out and evaluating all the incredible obstacles that this evolution has had to overcome to produce such an extraordinary universe? And are there other universes, an indefinite number of other universes, that have been less lucky and less successful, by way of compensating for our success? Above all, are there other universes significantly superior to our own?

Without becoming involved in the question of possible parallel universes, we can ask ourselves whether, in the only universe at our disposal, with its 100 billion galaxies, there are intelligences superior to our own. With SETI, based on our knowledge of bioastronomy, we, at the end of the 20th century, not only have the possibility but also the prospect of finding an answer. If so, perhaps we can learn something to our advantage. Answers to all these questions should enlighten us as to our place in the universe, and our likely future.

My feeling is that we are unlikely to be alone in this universe. This is why it is absolutely essential to carry out a search, and not close our eyes to such a wonderful adventure. The very recent acceptance of the paradigm of a biological universe can only serve to encourage us to follow this promising avenue of research. If we once manage to achieve the detection of an extraterrestrial civilization, a large number of others will follow in later years. We can then envisage forming part of a new community, an interstellar community, avid to meet one another, and we will be astonished that no one had searched for this newly found treasure much earlier and with far greater resolve.

Toward scientific recognition of SETI

The 500th Columbus Day, 12 October 1992, was chosen symbolically by NASA to inaugurate officially, on a national scale, the new SETI system that it had conceived, studied, promoted, defended, carried through, and financed, over the long period of 22 years. John Billingham, as Program Director, sent out invitations in the spring:

> The observations are scheduled to start during inauguration ceremonies, the Targeted Search in Puerto Rico at the Arecibo Telescope, the Sky Survey at JPL Goldstone Deep Space Communications Complex in California ... Frank Drake and Philip Morrison will be speaking at Arecibo, and Carl Sagan at Goldstone. The Project Engineering Teams will hand over the SETI signal detection systems to the Project Scientists. A few minutes later, Jill Tarter and Samuel Gulkis will simultaneously initiate the two searches formally ... In the first few minutes of observations, the data collected will exceed the sum of that from all previous searches.

It was a great moment. The action of turning those two small switches was comparable to Christopher Columbus's first steps on Guanahani, and Neil Armstrong's on the Moon in 1969. As yet land remains unseen: we do not know when the waves will reveal anything, nor what it will look like. In fact, SETI, like Columbus in his caravel, or Armstrong in his spacecraft, is just setting forth on a voyage into the unknown. We shall have to wait until we land to find out. If the faint beams of a lighthouse appear far off in the

213

depths of space, we shall have won. We shall know that there are other worlds, and that there are probably many of them; because once one is found, many more will be discovered. One of the best-known of the last century's leitmotifs, whose originator was Camille Flammarion, will take on a new significance: 'Is it not strange that the inhabitants of our planet should have almost all lived up to now without knowing where they are, and without suspecting the marvels to be found in the universe?'

Did we really think that after a few minutes listening, which in themselves amounted to more than that carried out over the last 32 years, a 'beep, beep' would be heard? No, the search was just beginning. Indeed, after a quarter of an hour the official guests began to become bored and the sophisticated computer was left alone to begin its long surveillance.

Now, under the aegis of the SETI Institute, Project Phoenix will take over. Regularly, over the weeks, months, and years, the computer will be examined, checked, and maintained, to keep its heart beating without pause at the rate of 130 million beats a second, which is why it was built. If an abnormal signal appears, it will sound the alarm all by itself, and that will give way to the pandemonium of action; the global network will be activated, doubtless only to confirm, quite frequently, that it was merely a false alarm.

When will the true signal be detected? Perhaps in a week, perhaps a century from now. In the meantime, scientists, engineers, and technicians will endeavour to build billions of channels, to detect even more complex signals, to explore other ranges of frequencies, and to develop new strategies. SETI ought to succeed!

The birth of SETI

The trail leading to that inauguration has been a long one! It has taken a whole generation. Yet the birth of SETI was sudden: at the beginning of the 1960s, Drake carried out his listening session; Cocconi and Morrison, and then Kardashev, published their papers. Drake and Bernard Oliver organized the first SETI conference under the auspices of the American Academy of Sciences, while the Soviet Academy of Sciences financed the first research. On the Soviet side, Shklovskii published an important book in 1962 on intelligence in the universe, and in 1964, Kardashev, Troitskii and a few colleagues organized their first SETI conference at the Buryakan Observatory in Armenia. The first international initiative was taken by the International Academy of Astronautics (IAA): one of its ten committees, the SETI Committee, was set up in 1966 by Rudolf Pesek, Professor of Aerodynamics at the Technical University in Prague. In 1968, John Billingham arrived at the Ames Research Center in California, after having invented the water-cooled underwear required for use with spacesuits. It is to him that we owe the creation, in 1970, of the first group set to study the detection of extraterrestrial signals. He appointed Bernard Oliver as head of the first great study, Project Cyclops, a technical masterpiece, which proposed the gradual construction of an immense group of 1000 radio telescopes, each 100 m in diameter.

Another important stage was an extremely small note lost in an enormous report. When starting out to open up a new field of exploration, it is often necessary to begin modestly, which may – provided the field is a valid one, of course – help it to become established. The report in question was the report of the meeting in 1979 of the World Administrative Radio Conference, which is charged with preparing, for the United Nations, rules for the use and allocation of radio wavelengths. The tiny note, number 722, said:

> In the bands [such and such], passive search is being conducted by some countries in a programme for the search for transmissions of intentional extra-terrestrial origin.

Then in 1982, Carl Sagan circulated a petition in scientific circles about extraterrestrial intelligence. Stressing the scientific merit, the benefits, and the interest in SETI, and the increasingly disturbing menace of radio pollution, he stressed the necessity of promptly undertaking such research, which is the only way of obtaining information about the existence of other civilizations in the universe. This petition carried the signatures of 70 scientists from a variety of disciplines, and from a large number of different countries.

The same year, another important step was taken. The International Astronomical Union (IAU), which brings together the 7000 professional astronomers in the world, set up its 51st Commission, entitled 'Bioastronomy, the search for extraterrestrial life.' Its aim is to organize research into life in the universe undertaken by astronomers, in accordance with seven main principles:

1. To search for planets in other stellar systems.
2. To study evolution of planets and their possibilities for life.
3. To detect extraterrestrial radio signals.
4. To investigate organic molecules in the universe.
5. To detect primitive biological activity.
6. To search for signs of advanced civilizations.
7. To collaborate with other international organizations, such as those devoted to biology, astronautics, etc.

This Commission immediately became one of the largest in the IAU, after those for galaxies and radio astronomy, with 300 members. It represents only about 4 % of professional astronomers, and the members are not all actively involved in bioastronomy. Nevertheless, the number of members bears witness to the keen interest in the investigation of extraterrestrial life. Half of the other commissions have interdisciplinary links with this research. This is why the seventh objective is important, because major breakthroughs frequently come from collaboration between different disciplines.

The 1980s were marked by numerous workshops and colloquia that brought together small groups of specialists. The general atmosphere was similar to that described earlier in connection with the Rosetta project (p. 38). When the IAU recognized bioastronomy, large international symposia began to be held, with greater structure and much more interdisciplinary in nature. They were a great success; the *Proceedings*, with details of the papers presented and the ensuing discussions, form the basic literature for this new discipline. The first was held in the United States, in Boston, and was entitled 'Recent developments'; the second, on 'The next stage', took place at Balaton, in Hungary, a sign of the longstanding interest in SETI in Eastern Europe. To maintain the balance, I proposed organizing the third in Western Europe, specifically in France. The theme was 'The exploration expands', and with the scientific co-organizer,

Mike J. Klein, former SETI Director at JPL and now involved with searching for planets, we tried to attract contributions reflecting the broadest possible range of interdisciplinary connections. The next symposium was held in California in 1993, thanks to a kind invitation from Frank Drake. Finally, John Billingham suggested that an IAA symposium should be held in 1995 on 'SETI and society', where, at the scientists' invitation, a hundred specialists would discuss cultural, political, legal, journalistic, religious, sociological, psychological, historical, literary, and artistic aspects of SETI. At my suggestion, France, which has a long and strong tradition of universal culture, will act as host at Chamonix-Mont Blanc to this very large meeting where the answers to the numerous questions that have already been raised in all these various disciplines will doubtless begin to germinate.

The International Space Year

Various initiatives were taken to mark the occasion of the 500th anniversary of Christopher Columbus's crossing of the Atlantic. Scientists around the world, through their international unions, decided that 1992 should be International Space Year, aimed at planning, coordinating, and undertaking a global investigation of our planet, using the powerful, space-borne, observational methods that have been developed in recent decades. We have to go back to 1957 to find a similar undertaking, when, during the International Geophysical Year, the world was also the subject of concerted, large-scale studies. But then it was without satellite resources, because it was in that year that Sputnik 1 became the first artificial satellite. As part of International Space Year, it was also decided to hold a World Space Congress, to bring together thousands of scientists, engineers, administrators, and decision-makers involved in space, for nine days of symposia, seminars, and meetings on the subject of discovery, exploration and cooperation. This vast undertaking, which took place in Washington, DC, became the responsibility of two large organizations: the Committee for Space Research (COSPAR) and the International Astronautical Federation (IAF), in cooperation with the International Academy of Astronautics and the International Institute of Space Law. A SETI Conference was held on this occasion, at which representatives of all the countries involved in the search were given the opportunity of presenting their work.

The results of this conference will be of great significance for our future. It will help us to protect our planet Earth and, above all, set us on the road toward a better understanding of the fragile and complex way in which its biosphere functions. We should not forget that although the human race does not yet have any control over the interior of the Earth, nor over deep space – which is probably lucky, given the initial mistakes that it generally makes – it does have fundamental effects on the biosphere, on which it, in turn, is utterly dependent. It is also worth remembering that when reference is made to the Earth as a spacecraft in which we are all traveling toward our common destiny, there is a tendency to think of it as a solid globe, 40 000 km in circumference. In fact, this is a misleading image, because basically we depend on an extremely thin, delicate, unstable, and still poorly understood component: the biosphere. This basically spherical shell surrounds the Earth, and reaches from the depths of the oceans, and from just a few decimeters beneath arable land on the continents (when there is any!), up through the troposphere to the lower stratosphere. When compared with the globe itself, the biosphere, within which every living thing exists, is only as thick as a skin 1 mm thick on a sphere 1 m across.

Although the World Space Congress should bring together essential and valuable elements involved in the protection of our environment, even more important, it should alert us to the urgent necessity of implementing such protection. Some of the safeguards and preventative measures that may emerge will be long-term projects for the 21st century, but others, such as those relating to the decrease in atmospheric ozone, ought to be implemented immediately.

Among all the symposia that were held during the World Space Congress, SETI was accorded a special place. This is a new and significant sign of recognition. What could be more natural, in fact, given that SETI is currently the ultimate expression of the great art of space exploration? In terms of distance, it will probably never compete with the records set by quasars, but with regard to the degree of understanding it could bring, with all the stars and galaxies scattered throughout the universe, it could reveal utterly unexpected high points. Can anyone imagine any celestial body, or any physical process, that is more complex or more highly developed than an advanced civilization?

SETI and Europe

Relative to the NASA giant, whose SETI budget grew from 2 million dollars a year to 14 million over six years, it might well be thought that the European contribution is very minor. This was true as far as money was concerned up to the time when NASA's SETI budget was withdrawn in 1994. Quite independently of that, however, and even excluding the former Soviet Union, scientific interest in SETI is extremely significant in Europe. The SETI Committee was created by a Czech, and originally the majority of its members were European. Currently Europeans amount to a quarter of the membership, and the vice-president is Hungarian. The Steering Committee of the IAA consists mainly of Europeans, and the editor-in-chief of *Acta Astronautica*, the journal set up by the IAA, is French. Of the 300 members of the IAU's Bioastronomy Commission, one quarter are European, and the Secretary is French. Four bioastronomy symposia have been held, one of which was in Hungary, and one in France. One quarter of the scientific papers on SETI are published by Europeans. The principal contributors are Austria, France, and Italy, followed by Germany, Hungary, the Netherlands, Poland, and the United Kingdom.

France was the only country actively engaged in searching for signals, using the radio telescope at Nançay, but Italy is joining in the great adventure, with the Large Northern Cross, at Bologna. This vast instrument, which is an interferometer consisting of a cylindrical paraboloid antenna, 534 m long and 35 m wide, together with 64 other cylindrical antennas, each 24 m by 8 m, will use the Earth's daily rotation to sweep the whole of the sky's northern hemisphere. Stelio Montebugnoli from the Radio Astronomy Institute of the Consiglio Nazionale di Ricerca [National Research Council; CNR] aims to connect a SETI data-acquisition system to the Large Cross. In addition, Christiano Batalli-Cosmovici, the Italian government's scientific advisor, a distinguished physicist, and ex-candidate astronaut, has set up a bioastronomy program within the CNR. These projects will perhaps allow Italy, whose involvement with space research dates back to the very early days, to play a greater role in the interesting efforts that Europe is undertaking in the SETI field.

France is not content to rest on its laurels. In 1989, for example, I proposed the idea of a global SETI network. If a candidate signal is detected, rapid communal action by people actually involved on a daily basis with SETI will be essential. The basic aim of the network would be to coordinate such work. My proposal was supported by the SETI Committee of the IAA and by the IAU's Bioastronomy Commission. Typical subjects that the network would have to consider concern SETI strategies, receivers, detectors, radio interference, logistics, tests in real time of the immediate verification of alerts and of the continuous monitoring of any potential signal, practice of immediate collective action that would have to be undertaken, and, finally, mutual communications by means of a suitably fast network. In principle, NASA would have had its own internal network, but despite the scale of its operation, it would still have been essential for it to be linked with the other SETI searches that are either in operation or in prospect in various parts of the world.

Given the significant contribution that it has already made over a large number of years, but which has been somewhat overshadowed, Europe ought to play an increasingly active role. France's commitment of the large Nançay radio telescope has set the pace. It ought not to remain an isolated example.

The Nançay radio telescope

Hidden right in the middle of a forest, in the region of Sologne 200 km south of Paris, are the giant lattice-work panels of the Nançay radio telescope. It is a forest of birches and pines, where the ground is covered with heather, with occasional secluded pools, and ancient manor houses hidden at the end of winding, shady tracks. It was chosen, a generation ago, as home to the newly born science of radio astronomy in France. Many of us enjoy leaving Meudon, which is a marvellous site in itself, during the weeks when we carry out our observing sessions at Nançay.

Under the farsighted guidance of Yves Rocard, Professor of Physics at the École normale supérieure, the then very young Jean-François Denisse (who soon became Director of the Paris Observatory and an expert in wave and plasma theory) and also of Jean-Louis Steinberg (the later pioneer of French space-borne radio astronomy) the École normale supérieure acquired a vast triangular tract of land in Sologne, of side 1.5 km, with the aim of creating an area free from radio interference and human activity. This isolation also reduced the cost of the operation: pioneers have to start small.

Toward the end of the 1950s, the first astrophysical successes were obtained, working with the 21-cm, neutral-hydrogen line. The pioneers from the École normale supérieure were taken under the wing of the Paris Observatory by its then Director, André Danjon, the leading light in French astronomy in the 20th century. They decided that France should have a giant radio telescope that could probe the world of the distant galaxies. In a broad clearing, 20 hectares in area, that had been cut in the southern portion of the reserve, they assembled, piecemeal, the vast metallic structure that they had devised, and slowly fitted more and more sensitive receivers.

On 15 May 1965, President Charles de Gaulle, welcomed by Denisse, officially inaugurated 'the largest radio telescope in the world.' Their first conversation revolved around SETI and the alleged detection of an extraterrestrial signal from CTA 102 that *Pravda* had announced on 14 April (p. 150). Denisse pointed out to General de Gaulle that it was, in fact, a completely natural 'message.' Immediately, the President decided to summon the Soviet ambassador, to show him that 'you can't pull the wool over our eyes in France.'

For more than a quarter of a century, the Nançay radio telescope's highly functional structures have remained standing, always looking impeccable in their silver paint, and as accurate as ever, thanks to ceaseless, careful maintenance. Its two vast wings simply invite the universe to communicate with Earthlings. The radio telescope was designed to capture the radio waves from any object between the southern horizon and the north celestial pole as it crosses the meridian. The radio waves first encounter a vast, fine-mesh rectangular panel, 200 m long by 40 m high, which can be rotated about a horizontal east–west axis. By changing the inclination of this reflector, which despite its size can be controlled to within a few millimeters, the radio waves can be directed toward a second, fixed reflector, 500 m south of it, that is 300 m long and 35 m high. This second reflector is curved, so that the radiation is concentrated at a single point, the focus, half-way between the two reflectors and a few meters above the ground. This is where the collecting system is located, and which basically consists of a metal horn, whose dimensions need to be extremely accurate, which directs the waves into a wave-guide, which again has to be very accurately made. This wave-guide finally ends at a dipole, where the weak electromagnetic fields that are received induce extremely weak currents, which are immediately amplified by the proper receivers. This system allows a large area of the sky to be observed, down to 40° south, while minimizing the number of moving parts in the vast reflecting surfaces, which amount to two hectares of mesh in total. In addition, the prime-focus horn is mounted on a mobile carriage, running on rails, which allows an object to be followed for an hour. The receiver is thus able to make long observations of a specific object, which in turn increases the sensitivity of the observations. Movement of the carriage is controlled by computer, ensuring perfect tracking of the object as it is carried westward by the Earth's rotation.

A collaboration between NASA and Nançay?

It was these features of the large radio telescope at Nançay that prompted me to suggest, at a Paris Observatory planning meeting in February 1985, a large-scale SETI collaboration, using the ten million channel system developed by NASA. A duplicate of the NASA system would be installed by the Americans at the Nançay radio telescope, in exchange for its communal use for SETI for a significant (later to be determined) period of time.

Second in the world only to the Arecibo radio telescope in terms of surface area at SETI's decimetric wavelengths, Nançay is an important trump card on the international scene. It was assessed favorably in 1987 by NASA, who declared that 'It is very suitable because of its large surface area, its vast coverage of the sky, and because of the possibilities it offers over a wide range of frequencies. It does, however, require improvements to make the most of the instrument: screens to protect against external interference, a more effective prime-focus system, and more sensitive receivers.'

In fact, I saw for myself in 1989 when visiting American institutes involved in SETI that the same problem of improving the prime-focus system also applied to the Arecibo telescope. I noticed much activity by a group of people beneath the hole in the center of the immense 300-m dish, and asked my guide to take me down there. It was a very strange feeling to be walking in the half-light underneath the vast inverted metal vault, where the ground is carpeted by the riotous vegetation that flourishes in the greenhouse effect that it creates. What I saw was a spherical cabin, containing two elliptical reflectors, which was about to be hoisted through the hole up to the focus, 150 m above our heads. The person in charge of this strange lifting operation was Lynn Baker, head of the Antenna Development Laboratory at Cornell University, where Arecibo's governing body is situated, at the National Astronomy and Ionospheric Center. He said 'It will be tested tomorrow ... It is a small-scale model for a new prime-focus system which will replace the old-fashioned, long "pencil", bristling with dipoles, that dates from the time the instrument was first built. By adjusting the shapes and positions of these secondary and tertiary reflectors, we shall be able to transform this old dish into a new–generation radio telescope, ten times as efficient.' The key advantages of this technological miracle is that the secondary may be adjusted so that no radiation from the ground outside the dish will enter the system – because the ground, at a temperature of 300 K, emits a vast flux of 'dazzling' radiation – and adjustments to the tertiary ensure that the radio emission from the object being investigated is picked up as efficiently as possible after being reflected by the dish.

As in many other old systems, the edges and the center are difficult to use, giving a corresponding reduction in the effective surface area. In addition, the cabin will prevent radio interference from directly reaching the focus. Finally, after being reflected backward and forward between the secondary and tertiary reflectors, the radiation will dive into a feed horn, located close by, and thus well-protected from interference, from which it will reach the receiver.

Could this 'magic potion' of a double-reflector system be applied to Nançay? It too would then become a new-generation radio telescope, and could continue to serve radio astronomy for another twenty-odd years! According to Lynn Baker, *a priori* the Nançay case was more complicated than at Arecibo, because its collecting surface was extremely elongated, rather than circular. However his initial calculations, together with others by Sebastian von Hoërner, one of radio astronomy's high priests, have shown that Baker's system would be suitable at Nançay. Since then, the members of the telescope's Antenna Group, headed by Gabriel Bourgois at Meudon, have tackled the difficult problem of a double reflector. In 1992, a Phase A study was begun. This phase should enable the size of the reflectors to be defined, which is a decisive factor in the final cost.

At the end of 1990, in preparation for a possible future installation, NASA sent us two of its systems to evaluate the background level of interference, which is a major problem for sensitive observations. Incredible! They showed that Nançay, along with Green Bank in West Virginia, was one of the two best sites in the world. This result must be because our telescope has a very low profile, and is sited at ground level, hidden away in the center of a vast flat forest that helps to screen it from interference, and in a region with little industrialization. Encouraged by these results, Éric Gérard from Meudon, an inveterate foe of interference, and Bernard Darchy the receiver engineer at Nançay, have been lobbying the regional government to declare the area a radio-interference free zone. They have also begun investigations of a protective mesh fence that would surround the whole instrument, rather like the wire around a vast tennis court.

Despite the Congressional killing of all NASA funding for SETI, the SETI Institute is continuing with the targeted search, and hopes, after observational runs at Parkes in Australia, to be able to come to Nançay with its ten million channel receiver for the final years of this century.

In addition, I have started unofficial contacts with the Director of the Berkeley SERENDIP project, Stuart Bowyer, and the Chief Project Scientist, Dan Wertheimer. The idea of installing their 160 million channel analyzer as a piggyback addition to the large French radio telescope, and using it for common searches, was very well received. The fate of such a bright prospect depends upon funding from the University of California and from the Paris Observatory, or from private donations.

Bibliography

GENERAL WORKS

Drake, F. D. and Sobel, D., 1992: *Is Anyone Out There?: The Scientific Search for Extraterrestrial Intelligence*, Delacorte Press, New York.

PROCEEDINGS OF BIOASTRONOMY SYMPOSIUMS

Papagiannis, M. D. (ed.), 1984: *The Search for Extra-Terrestrial Life: Recent Developments*, International Astronomical Union Symposium No. 112 (Boston, USA), Reidel, Dordrecht.

Marx, G. (ed.), 1987: *Bioastronomy, The Next Steps*, International Astronomical Union Colloquium No. 99 (Balaton, Hungary), Kluwer, Dordrecht.

Heidmann, J. and Klein, M. J. (eds.), 1990: *The Search for Extra-Terrestrial Life: The Exploration Broadens*, Third International Bioastronomy Symposium (Val Cenis, France), *Lecture Notes in Physics, No. 390*, Springer-Verlag, Heidelberg.

Shostak, G. S. (ed.), 1995: *Progress in the Search for Extraterrestrial Life*, 1993 Bioastronomy Symposium, Santa Cruz (USA); Astronomical Society of the Pacific Conference Series, Vol. 74.

PROCEEDINGS OF SETI FORUMS HELD BY THE INTERNATIONAL ASTRONAUTICAL ACADEMY

Seeger, C. L. and Martin, A. R. (eds.), 1989: The Search for Extra-Terrestrial Intelligence, *Acta Astronautica*, special issue, Vol. 11, No. 11, Pergamon Press, Oxford.

Tarter, J. C. and Michaud, M. A. (eds.), 1990: SETI Post Detection Protocol, *Acta Astronautica*, special issue, Vol. 21, No. 2, Pergamon Press, Oxford.

Heidmann, J. (ed.), 1992:   SETI 3, The Search for Extra-Terrestrial Intelligence, *Acta Astronautica*, special issue, Vol. **26**, No. 3/4, Pergamon Press, Oxford.

Index